Taiwan's Offshore Islands

Pathway or Barrier?

Bruce A. Elleman

NAVAL WAR COLLEGE PRESS
Newport, Rhode Island

To Pat, who always kept my spirits high, even in my darkest hours

In Chinese, the number 4 sounds like the word for "death" and so is unlucky.
The number of this monograph in the Newport Papers series—44, which sounds like
"dying and dead"—is very unlucky indeed. Each chapter begins
with a Chinese proverb related to death or war.

Library of Congress Cataloging-in-Publication Data

Names: Elleman, Bruce A., 1959– author. | Naval War College (U.S.). Press, publisher. | Naval War College
   (U.S.). Center for Naval Warfare Studies, issuing body.
Title: Taiwan's offshore islands : pathway or barrier? / Bruce A. Elleman.
Other titles: Newport paper ; no. 44. 1544-6824
Description: 1st edition. | Newport, Rhode Island : Naval War College Press, 2019. | Series: Newport paper,
   ISSN 1544-6824 ; 44 | "Center for Naval Warfare Studies." | Includes bibliographical references and index.
Identifiers: LCCN 2019002267 | ISBN 9781935352693 (paperback)
Subjects: LCSH: Taiwan—Foreign relations—China. | China—Foreign relations—Taiwan. | United States—Foreign
   relations—Taiwan. | Taiwan—Foreign relations—United States. | Taiwan Strait—Case studies. | Taiwan Strait—
   Strategic aspects. | Islands—Taiwan—Strategic aspects. | Cold War.
Classification: LCC DS799.63.C6 E44 2019 | DDC 327.51249051—dc23 | SUDOC D 208.212:44
LC record available at https://lccn.loc.gov/2019002267

Naval War College

Newport, Rhode Island
Center for Naval Warfare Studies
Newport Paper Forty-Four
January 2019

President, Naval War College
Rear Adm. Jeffrey A. Harley, USN

Provost
Dr. Lewis M. Duncan

Dean of Naval Warfare Studies
Thomas J. Culora

Naval War College Press

Director: Dr. Carnes Lord
Managing Editor: Robert C. Ayer

Telephone: 401.841.2236
Fax: 401.841.1071
DSN exchange: 841
E-mail: press@usnwc.edu
Web: usnwc.edu/Publications/Naval-War
-College-Press
Twitter: http://twitter.com/NavalWarCollege

Printed in the United States of America

The Newport Papers are extended research projects that
the Director, the Dean of Naval Warfare Studies, and the
President of the Naval War College consider of particular
interest to policy makers, scholars, and analysts.

The views expressed in the Newport Papers are those of
the authors and do not necessarily reflect the opinions of
the Naval War College or the Department of the Navy.

Correspondence concerning the Newport Papers may be
addressed to the Director of the Naval War College Press.
To request additional copies, back copies, or subscriptions
to the series, please either write the President (Code 32S),
Naval War College, 686 Cushing Road, Newport, RI
02841-1207, or contact the Press staff at the telephone, fax,
or e-mail addresses given.

ISSN 1544-6824

ISBN 978-1-935352-69-3

Contents

Foreword

Following the Nationalist defeat on the mainland in 1949, Chiang Kai-shek and his followers retreated to Taiwan, relocating the Republic of China (ROC). To many it seemed almost certain that the People's Republic of China (PRC) would attack and take Taiwan, perhaps as early as the summer of 1950. Control over a number of offshore islands, especially Quemoy (Jinmen) and Matsu (Mazu), became a deciding factor in whether the PRC could invade Taiwan or, conversely, the ROC could invade the mainland. Twice in the 1950s tensions peaked: during the first (1954–55) and second (1958) Taiwan Strait crises. During both events the U.S. government intervened diplomatically and militarily. This study will discuss the origins of these conflicts, the military aspects of the confrontations, and in particular the complicated and largely secret diplomatic negotiations—including two previously unknown Eisenhower-Chiang secret agreements—going on behind the scenes between the U.S. government and the Nationalist government on Taiwan.

BRUCE A. ELLEMAN
William V. Pratt Professor of International History
Naval War College

Acknowledgments

The author would like to thank those who shared their insights and expertise, including the archivists, archival technicians, and archival specialists at the Harry S. Truman Presidential Library and Museum, including Jim Armistead, David Clark, Janice Davis, Sam Rushay, Randy Sowell, Tammy Williams, and Kurt Graham. At the Dwight D. Eisenhower Presidential Library, Kevin M. Bailey was extremely helpful, as was Michelle Kopfer. For in-depth information about the Taiwan Patrol Force, I am indebted to James Barber, Lyle Bien, Bob Chamberlin, Doug Hatfield, Jay Pryor, Paul Romanski, and Michael Westmoreland. At the Naval War College, I benefited from the support of Timothy Schultz, Phil Haun, Lewis Duncan, Tamara Graham, and Jeffrey Harley. At the Naval History and Heritage Command, John Hodges gave me incredible access to the U.S. Navy archives. At the Library of Congress, College Park, Maryland, Don Mcilwain and Alan Lipton were very helpful. Ben Primer helped me use the John Foster Dulles Papers at the Seeley G. Mudd Manuscript Library, Princeton University. I also owe a considerable debt to the Naval War College Library. At the Naval War College Press, I would like to thank Carnes Lord, Robert Ayer, and Pel Boyer. I am especially indebted to Andrew Marshall, director of the Office of Net Assessment until his recent retirement, and Andrew May, deputy director, for their ongoing support for this and other China-related projects. All American and British sources with classification designations have been declassified in accordance with standard procedures of the respective countries.

Naval Terms and Acronyms

C	**CHICOM***	Chinese Communist
	Chinat*	Chinese Nationalist
	CHINCOM*	China Committee
	CIA	Central Intelligence Agency
	CinCPac	Commander in Chief, U.S. Pacific Command
	CNO	Chief of Naval Operations
	COCOM	Coordinating Committee for Multilateral Export Controls
F	**FRUS**	Foreign Relations of the United States
G	**Gimo**	Generalissimo
	GPO	U.S. Government Printing Office
	GRC	Government of the Republic of China
J	**JCS**	Joint Chiefs of Staff
L	**LCM**	landing craft, mechanized
	LCVP	landing craft, vehicle, personnel
M	**MAAG**	Military Assistance Advisory Group
N	**NHHC**	Naval History and Heritage Command
	NIOHP	Naval Institute Oral History Program
	NSC	National Security Council
O	**ONI**	Office of Naval Intelligence

P	**PLA**	People's Liberation Army
	PLAAF	People's Liberation Army Air Force
	PLAN	People's Liberation Army Navy
	PRC	People's Republic of China
R	**ROC**	Republic of China
T	**TNA**	The National Archives (of the United Kingdom, Kew)
U	**USIA**	U.S. Information Agency
	USSR	Union of Soviet Socialist Republics (the Soviet Union)
W	**WEI**	Western Enterprises Incorporated

* Variously capitalized in the literature on the subject.

CHAPTER ONE

The Two Chinas and the Battle for Control of Offshore Islands

War is death's feast. (战争是死神的筵席.)

This Newport Paper will examine the role of offshore islands in twentieth-century East Asian history, in particular those islands in the Taiwan Strait that were disputed by the People's Republic of China (PRC) and the Republic of China (ROC) during the 1950s and afterward, and how these apparently insignificant islands impacted Cold War history. The history of the 1945–91 Cold War can be divided into three periods: breaking apart the Sino-Soviet monolith, forging Sino-American cooperation against the Soviet Union (or USSR), and the two-front destruction of the USSR. The Harry S. Truman Presidential Library and the Dwight D. Eisenhower Presidential Library contain especially pertinent information on the first of these three periods, allowing a greater understanding of how and why Washington adopted policies with the strategic goal of breaking apart the Sino-Soviet alliance.

For much of this period, from 1950 to 1979, the U.S. Navy conducted a buffer operation in the Taiwan Strait. The most important goal of this operation was to neutralize the region—to ensure that neither side attacked the other and so perhaps brought on global war. While it is often assumed that the U.S. government's focus was just the Taiwan Strait, in fact there were equally valid concerns that after "China [was] Communized, the remainder of Asia and the islands to the south and east [would] be vulnerable to rapid Communization." Owing to East Asia's geography, "there are no natural barriers to the advance of Communism from China to Indo-China, Siam, Burma and thence to Malaya, India, Indonesia and the other islands of the Western Pacific."[1] Because of the enormous Nationalist losses in China's civil war, by late 1949, when the Nationalists were driven off the mainland to Taiwan, America's defense line had moved from China's Pacific coast to the various islands of the Aleutians and from there down through the Kurils, Hokkaido, the main Japanese islands, Okinawa, Formosa, and the Philippines.

Formosa, in particular, was "in the direct line—in fact, it is the center of our whole defense."[2] Seen in this perspective, the defense of Taiwan was essential to the defense of all of East Asia.

To set the stage, this chapter will examine the strategic importance of the Taiwan Strait in the creation of the "two Chinas" conflict, the adoption by the Nationalists of a naval blockade against the PRC, and their cooperation with anticommunist guerrillas on the offshore islands. It will also examine in greater detail the location of the various offshore islands, their military or strategic importance, and factors that led to their retention by the Nationalists or to their return—either by force or after an orderly withdrawal—to the control of the PRC.

Historical Background

The collection of offshore islands between mainland China and Taiwan has often played an important military role.[3] The Taiwan Strait is a very strategic region, since it lies along the primary north–south sea-lane in East Asia. Japanese, Korean, and northern and central Chinese luxury goods and produce flowing primarily from north to south must transit this strait to reach Southeast Asia, as must luxury goods and raw materials flowing from south to north. The Taiwan Strait has long been a choke point, and which-ever of the two competing regimes dominated both sides could close it to international shipping. Such an action would force commercial vessels to take the longer and more exposed route to the east of Taiwan.

Taiwan has accordingly been fought over many times, including in the seventeenth cen-tury between Ming loyalists and the Manchus, who had displaced the Ming dynasty in 1644; in the eighteenth century between the Manchus and local rebels; and during the 1880s' Sino-French War. In the 1895 Treaty of Shimonoseki that ended the first Sino-Japanese War, China ceded Taiwan to Japan in perpetuity. Japan maintained sovereign-ty over Taiwan for fifty years, until 1945, at which point, according to the terms of the Cairo and Potsdam agreements, Taiwan was returned to "China"—which at this stage meant "Nationalist China," the internationally recognized government.

The division of China into a Taiwan-based Republic of China and the mainland-based People's Republic of China was a direct outcome of World War II and the Chinese civil war. During World War II, the U.S. government encouraged the Nationalists and Com-munists to form a coalition government. As Nationalist rule imploded, the Communists orchestrated an increasingly effective campaign to rally popular backing. It looked like the USSR might attack in their support. President Truman decided it was not worth the billions of dollars and millions of men it would take to defend China. On July 26, 1946, Fleet Adm. William D. Leahy, chairman of the Joint Chiefs of Staff (JCS), duly reported to Truman that if there was a Soviet attack all U.S. troops in China would immediately withdraw "southward" to be evacuated through the city of Tsingtao (Qingdao, in Shandong Province).[4]

The U.S. government refused to get involved in the Chinese civil war. As early as November 14, 1945, Congressman Ellis E. Patterson wrote to Truman condemning "intervention on our part" on the side of any Chinese faction.[5] Truman assigned Gen. George C. Marshall, his special envoy to China, to find a solution, telling Secretary of Commerce Henry A. Wallace on December 18 that he had instructed Marshall to use American influence "to the end that the unification of China by peaceful democratic methods may be achieved as soon as possible." Marshall, for his part, reported back that "a China disunited and torn by civil strife could not be considered realistically as a proper place for American assistance," whether technical, military, or economic. Truman then ordered Wallace to suspend all ongoing talks with Chinese officials that "might encourage the Chinese to hope that this Government is contemplating the extension of any type of assistance to China, except in accordance with the recommendations of General Marshall."[6] The Marshall mission would be widely considered a failure, but as Maurice Votaw, an American academic who had advised the Nationalists, would point out, Marshall had had "all the cards stacked against him in advance, so . . . could not have possibly brought about a settlement."[7]

On June 2, 1947, the Department of State warned that Nationalist troop morale in Manchuria was so poor that "there is a possibility of a sudden military debacle which would lay all Manchuria open to the Communists."[8] During September 1947, in fact, the Communists were able to shift to the offensive in Manchuria. They pushed the retreating Nationalist troops into a small triangle bounded by the cities of Jinzhou, Changchun, and Mukden.

By May 1948, the People's Liberation Army (PLA) had cleared most of the Nationalist troops from Manchuria. That November, the Communists wiped out another hundred thousand Nationalist troops. Finally, on December 15, 1948, after sixty-three days of fighting, the Communists took Xuzhou, opening the road south to the Yangzi River and to Nanjing.[9] The U.S. Army had at that point a small number of troops—sixty officers and four thousand men—in Qingdao, and during November 1948 the Joint Chiefs had begun seriously to discuss withdrawing them.[10] American military advisers in China were "unanimous in the view that short of the actual employment of US troops in China no amount of military assistance can now save the Chiang Kai-shek regime in the face of the present political, military and economic deterioration."[11] Even Chiang's personal secretary called the defeat in Manchuria a "shameless debacle."[12]

But although they had lost North China, the Nationalists retained their traditional power base in South China. Moreover, as early as November 29, 1948, no less a personage than the mayor of Boston, James M. Curley, had suggested to Truman that "our most important ally in the event we should have trouble with Russia should prove [to be] China, with its tremendous natural resources and enormous population[;] . . . the powerhouse of Russia's war industries is located in the vicinity of Siberia and the nearest approach to Siberia is from China."[13] A year later, Frederick C. McKee argued, "China is the key to the defense of Asia. . . . [I]f we help the Chinese to hold West China, its airfields could be our closest bases for counter attack on the Soviet Atom Bomb plants North of Lake Baikal."[14]

On January 6, 1949, members of the National Assembly of the Republic of China begged Truman for assistance in their "campaign against totalitarian rule in defense of freedom and democracy."[15] On February 9, 1948, William C. Bullitt, a former American ambassador to Moscow, accused Truman and Marshall of not doing enough to stop the Communists: "If Manchuria falls to the Communists, their sweep south to the Yangtze River will be rapid, and they may then be able to conquer all China; and you, Mr. President, will, thereafter, have to account to the American people for the consequences to them and to the world of control of China by [Joseph] Stalin [the Soviet dictator]."[16] In fact, Bullitt was convinced, "you and your Secretary of State, General Marshall, would go down in history as having been unintentional patrons of the Soviet campaign for destruction of the United States and domination of the world."[17] Maj. Gen. Claire Lee Chennault agreed, warning the White House on June 10, 1949, "that all of South East Asia will rapidly go Communist once China has fallen."[18]

Nevertheless, on August 3, 1948, a Central Intelligence Agency (CIA) assessment of the Nationalists concluded that "Chiang's position is steadily deteriorating, and his Government is in such [a] precarious situation that its collapse or overthrow could occur at any time."[19] Four months later, an update reiterated that Chiang might be "removed from his position as President of the Republic of China at any time," adding that "his removal would be brought about by forced resignation or a coup d'etat, possibly including his assassination."[20] According to John Leighton Stuart, American ambassador to China since 1946, "any effort to keep Chiang in power through American aid would not only be undemocratic but would also arouse greater sympathy for the Communist cause and create violent anti-American feeling."[21]

Many foreign commentators assumed that China would now be divided into a north and south. For example, Irving Zuckerman advised Clark Clifford at the White House on November 30, 1948, that "it is quite possible that peace in China may be almost immediately possible by *letting the Red Chinese have the northern part of China*

(Manchuria), with Chiang having the southern part of China," a region the Nationalists would find "easier to administer."[22] Ambassador Stuart effectively concurred, projecting on November 26, 1948, that the Nationalists would make an "all-out effort to contain the Communists north of the Yangtze River."[23] The CIA assessed that if the Nationalists relocated to South China, American "aid in the form of capital, food, equipment, and intelligent direction could probably give the National Government [a] reasonable chance of carrying on effective anti-Communist resistance from South China," although there would be "little future safeguard from eventual defeat and collapse."[24]

To many, then, the Yangzi River could act as the new boundary, at least temporarily dividing China equally between a Communist north and a Nationalist south. Stalin had reportedly ordered Mao not to cross the Yangzi River;[25] the U.S. embassy explained on January 31, 1949, that "the Kremlin planners are fearful that the Communists may be able to develop too strong and independent a state."[26] But on February 25, 1949, the Nationalist navy's flagship, *Chongqing* (the former light cruiser HMS *Aurora*), mutinied. This event was a symbol of the waning Nationalist mandate to rule, and Secretary of State Dean Acheson referred to it at a National Security Council (NSC) meeting in those terms: "The defection to the Communists of the only cruiser in the Chinese fleet is symptomatic of the uncertainties in the present situation, particularly with respect to political conspiracy and 'deals' in the higher echelons."[27] By the end of April 1949 much of the rest of the Nationalist fleet guarding the Yangzi River had also defected. On April 20, 1949, Communist forces crossed the Yangzi and three days later overran Nanjing. Thereafter, the PLA quickly consolidated control over all of mainland China, taking Shanghai and Wuhan in May, Xi'an and Changsha in August, Guangzhou in October, and the Nationalist wartime capital of Chongqing in November.[28]

The PLA rapid advance southward forced the remaining Nationalist units to retreat to Taiwan. According to a CIA analysis of October 19, 1949, the 681,000 Nationalist combat forces had lost the "will to fight," for six major reasons:

> (1) Army politics, which keep incompetent officers in positions of high command; (2) general professional incompetence in the armed forces; (3) the personal interference of Chiang Kai-shek in tactical operations; (4) Chiang's refusal to supply money and materiel to commanders on the mainland; (5) inadequate pay, food, clothing, and equipment for the troops; and (6) chronic graft and corruption practiced by senior officers at the expense of the men.[29]

By contrast, the PLA experienced enormous growth, from an estimated half a million troops in mid-1945 to 1.3 million in mid-1946, two million in mid-1947, 2.8 million by mid-1948, and four million by early 1949.[30] American officials at this time considered the Chinese communists to be little better than Soviet puppets; according to Secretary of State Dean Acheson, "This Chinese Government is really a tool of Russian imperialism in China." The most important thing to avoid was adopting "any course of action that would solidify the Chinese people behind the Communist regime."[31] Later, Acheson predicted that "conflict would eventually develop between China and the USSR" and argued that "we should seek to take advantage of this conflict when it developed and meanwhile avoid actions which would deflect Chinese xenophobia from Russia to ourselves."[32]

Chiang Kai-shek and his advisers officially moved the Nationalist government to Taiwan on December 8, 1949, the eighth anniversary of the Japanese attack on Pearl Harbor, to continue their anticommunist struggle.[33] Even then, Chiang claimed that the ROC regime remained the legitimate government of all of China. Meanwhile, in late September 1949, Mao Zedong assembled a new Political Consultative Conference that elected him chairman of the central government and made Beijing the capital once again. On October 1, 1949, Mao officially proclaimed the creation of the People's Republic of China. Concerned about the strategic weakness of its colony Hong Kong, Great Britain, the main U.S. ally in Europe, extended official recognition to the PRC government on January 6, 1950.

During the summer and fall of 1949, Nationalist forces fiercely defended the numerous offshore islands that they held. The Nationalists initially kept one regiment of marines on the Miao Islands, north of the Shandong Peninsula, to blockade the Bo Hai Gulf and the northern ports and fortified Zhoushan and the Saddle Islands to blockade the Yangzi River. Meanwhile, the islands of the Tachens (Dachens), Matsu (Mazu—actually a group of thirty-six islets), Quemoy (Jinmen), and the Pescadores (Penghu) near Taiwan; Lema and Wanshan Islands near Guangzhou; and Hainan Island, fifteen miles off the southern coast—all were used after 1949 by the Nationalist government in Taiwan to conduct a naval blockade of the PRC.

The Nationalist Blockade Strategy

The Nationalists had been forced to retreat but they were not defeated. Chiang simply shifted from a land-based offensive to a naval one, a blockade. The ROC had already agreed to purchase "all surplus property owned by the United States in China and on seventeen Pacific Islands and bases";[34] it was charged only $95,500,000 for equipment valued at $1,079,300,000.[35] The Nationalists controlled a total of 824 vessels of various types. By 1947, the Nationalist navy had grown to almost forty thousand men and by October 31, 1948, to 40,859, including 8,062 officers.[36] By late 1949, however, the Nationalist navy, though still comparatively large, had dropped to an estimated thirty thousand men and 150 ships, and morale was low; the CIA considered it "possible that many units might defect."[37] Nevertheless, the Nationalists pursued their blockade against the PRC, working with a number of autonomous guerrilla movements on islands off China's coast. Later, the U.S. Navy provided military assistance—especially aircraft—that made air patrols of the blockade possible. The Nationalist blockade lasted from 1949 through 1958.[38]

In 1947, its navy both equipped and manned, the Nationalists had attempted to halt Soviet shipments to the Chinese communists in Manchuria via Port Arthur (now Lüshun) and Dalian by naval blockade, albeit mainly a paper one. Because Manchuria's ports were already closed to most foreign shippers, the blockade of Lüshun and Dalian elicited no complaints from the foreign powers. But on June 18, 1949, the National-ists announced that any Chinese port not under their control would be closed to trade as of midnight on the 25th. The next month, however, the Admiralty's Intelligence Division warned that if the Nationalists lost the Miao Islands and the entire Zhoushan area including Dinghai, as was then likely, the blockade would have no effect north of the Yangzi River, only a "limited and sporadic" one on trade entering and leaving the Yangzi River, in which case the "Nationalist blockade could only be effective south of Fuzhou."[39]

As the Nationalists retreated from their northern bases, therefore, the focus of the naval blockade moved southward. The Nationalist retreat to Taiwan was a major maritime undertaking, during which the Nationalist navy and other ships pressed into service transported approximately two million civilians and six hundred thousand soldiers across the strait. On August 6, 1948, the CIA argued that American bases on the Ryukyu (Okinawa) Islands would "give the US a position from which to operate in defense of an unarmed post-treaty Japan and US bases in the Philippines and other Pacific Islands."[40] Taiwan was a key part of the island chain running from the Aleutians through Hokkaido to Japan and Okinawa to the Philippines. If it fell, the PLA could use it as a base from which to invade other islands in the chain, as well as to interfere with international shipping.

On March 14, 1949, the CIA suggested that supporting Taiwan might strengthen the "will to resist Communism in Japan, in Korea, in the Philippines and elsewhere throughout the Far East," particularly if an American aid program "were developed in such a way as to secure local stability and contentment in Taiwan."[41] On October 26, 1949, the NSC formally included Taiwan in what it called America's "first line of strategic defense" in East Asia, arguing that this "minimum position will permit control of the main lines of communication necessary to United States strategic development of the important sections of the Asian area."[42] On November 21 the Department of State opined that "the resources of that island [Taiwan] appear adequate, if resolute steps are taken to utilize them effectively, to achieve our objective of denying the island to the Communists by political and economic means." However, what Taiwan really needed was a "spiritual regeneration," one that would produce new leaders able to "devote them-selves with determination to their cause and which would revitalize their followers."[43]

Gen. Omar N. Bradley, chairman of the Joint Chiefs of Staff, vocally supported protecting Taiwan. On July 27, 1950, he forwarded to the secretary of defense a four-page
"Memorandum on Formosa" that had been written by Douglas MacArthur on June 14.
The paper urged the importance of the Pacific island chain: "Geographically and strategically Formosa is an integral part of this offshore position which in the event of hostilities can exercise a decisive degree of control of military operations along the periphery
of Eastern Asia." In particular, during times of conflict Taiwan's position would give
U.S. forces the "capability to interdict the limited means of communication available
to the Communists and deny or materially reduce the ability of the USSR to exploit the
natural resources of East and Southeast Asia." Calling Taiwan an "unsinkable aircraft
carrier and submarine tender," MacArthur emphasized that the island was at "the very
center of that portion of our [defensive] position now keyed to Japan, Okinawa, and the
Philippines."[44]

During October 1949 a Communist attack on the Nationalist-held base on Jinmen,
bitterly opposed by Nationalist troops, failed: "The [ROC] armored division, which
is under the Generalissimo's son [Chiang's adopted son, Chiang Wei-kuo], is doing a
splendid job, as was shown in the battle of Kingman [sic] [Jinmen] with a loss of approximately 10,000 Communist soldiers and an additional 12,000 Communists captured."[45]
However, a number of strategic islands in the north were lost, effectively narrowing
the blockade to central and southern China. During 1950, the PLA advanced toward
the southern coastline, where, in spite of naval and air inferiority, it overwhelmed the
Nationalist base on Hainan Island in February–May, the Zhoushan Archipelago in May,
and Tatan Island in July. By the summer of 1950, therefore, the Nationalists had lost
their crucial island bases in the Bo Hai Gulf, off the mouth of the Yangzi River, and on
Hainan. These losses cut the Nationalist blockade's reach by over half.

Traditionally, offshore islands had acted as "stepping-stones" to Taiwan for invaders.
Thus, Nationalist control of the offshore islands was critical to deter the Communists.
By the spring of 1950, nevertheless, Communist forces seemed to be ready to invade.
April and May were considered to be the best months to attack, since typhoons were un-
usual then.[46] The PLA concentrated thousands of junks in the port cities along the Taiwan Strait in preparation for a massive amphibious invasion.[47] The Communists could
assemble "7,000 ships and craft" and "200 aircraft" to transport two hundred thousand
troops across the strait.[48] To foil such an attempt the Nationalists relied on their control
over the large island bases of Jinmen and Mazu, right off the coast of Fujian Province,
and over the Penghu Islands in the strait, halfway between the mainland and Taiwan.
The Nationalists turned to "guerrillas" on islands along the southeastern two-thirds of
China's coastline to maintain the blockade.

Nationalist-Supported Guerrilla Movements

Rather than fortifying all of these offshore islands themselves, the Nationalists worked with groups of anticommunist guerrillas. The degree to which the Nationalists interacted with and controlled their guerrilla allies was intentionally left unclear. For example, in the Dachens, naval interdiction was carried out only by guerrillas, which allowed the Nationalists to avoid blame. To motivate the guerrillas, it was decided early on that their crews would share in the prize money from captured blockade-running ships.

Not surprisingly, the Nationalist blockade in distant locations such as the Dachen Islands largely depended on guerrillas. The Nationalists portrayed them as anticommunist fighters. But to many foreign shippers, the pro-Nationalist guerrillas appeared little better than modern-day pirates. As reported by one British official in 1952, "Indications are that most of the piracies are done by Nationalist guerrillas not under the effective control of Taipei. The border line is obscure. In 1948 it was respectable for bandits to masquerade as Communist Liberation forces; in 1952 it is respectable for pirates to masquerade as the Nationalist Navy."[49] Unlike regular Nationalist troops, these guerrillas were primarily interested in their own financial gain—from either taking bribes or confiscating the foreign shipments. As the U.S. Navy's Office of Naval Intelligence (ONI) reported, "Actually, even though the Nationalists have the capability, they probably do not wish to suppress the guerrilla activities because it would undoubtedly lose them the guerrilla support in the coastal islands."[50]

In mid-January 1951, following China's intervention in the Korean War, the NSC recommended for the first time that the U.S. government "furnish now all practicable covert aid to effective anti-communist guerrilla forces in China."[51] American assistance to the Chinese guerrillas began that month. The NSC estimated that there were seven hundred thousand guerrillas in China, half of them owing allegiance to the Nationalists. To support bands in China's interior would require airdrops, with a high likelihood that many supplies would be lost. Supplying bands along the coast, in contrast, was "relatively simple," and the benefits of doing so were potentially great: "Successful guerrilla action should eliminate the physical aspects of communist support in Indochina, occupy the attention within China of the major part of the [Chinese communist forces, or PLA], and counter the myth of communist invincibility." While the Nationalist troops on Taiwan were numerous, this report concluded, "the optimum use of the Nationalist forces under present conditions appears to be in support of guerrilla operations on the mainland."[52]

On Lower Dachen Island, the CIA-sponsored Western Enterprises Incorporated (WEI) sent American advisers to train the guerrillas. The idea of a Military Assistance Advisory Group (MAAG) in China dated from February 26, 1946;[53] on July 16, Truman had authorized the U.S. Navy to "detail not more than one hundred officers and two hundred enlisted men of the United States Navy or Marine Corps to assist the Republic of China in naval matters."[54] By 1949, however, civil war had led to greater chaos in China, and the NSC warned that "units of the U.S. fleet should not now be stationed at or off Formosan ports in support of the political and economic measures" on Taiwan.[55]

Because of that chaos, American plans to support the Nationalists did not go fully into effect on Taiwan until the early 1950s. Once the Nationalists lost the mainland, they asked Washington for permission to "employ as a matter of immediate urgency a small group of American advisers composed of civilian, economic, industrial, agricultural and military advisers." This group of 130–50 would be used to build up defenses to save Formosa from invasion and thereby "greatly strengthen the resistance forces on the mainland [and] would give new courage to millions of Chinese who are opposed to communist domination and control."[56]

By 1951 the number of American advisers on Formosa totaled about 650, half with the MAAG, the other half with a variety of smaller organizations. One, WEI, appeared to be engaged in "guerrilla training and psychological warfare."[57] At any one time, about half the advisers—from 250 to three hundred—were assisting the Nationalist navy. Lon Redman, a WEI adviser, attempted to bulldoze an airstrip on the Dachens to handle C-46 cargo planes, but this plan had to be abandoned once it became clear that he "could not clear a strip long enough to permit a C-46 to land unless he bulldozed down almost to sea level."[58]

Contested Control over Offshore Islands

By 1953, the Nationalists controlled twenty-five offshore islands. On July 30, 1953, a U.S. Navy report, "Security of Offshore Islands Presently Held by the Nationalist Government of the Republic of China," divided twenty of these into three categories. In Category I were four islands off Fuzhou, including Mazu, and four off Xiamen, including Jinmen, that "could be used to counter Chinese Communist invasion operations."[59] Retaining these eight islands was considered militarily desirable for defensive purposes; Xiamen and Fuzhou in particular were "natural harbors for the staging of a sea invasion of Formosa." If these islands were seized by the PRC, however, their "function could largely be replaced by an increased use of naval craft which could blockade these two harbors."[60]

The Nationalists also hoped to use these two islands to return to the mainland. But the region of China opposite them was mountainous and had "few means of communication" with the interior of China, so a Nationalist invasion south of Xiamen or north of Fuzhou "would be more appropriate."[61] In any case, defending the Category I islands "minimizes rather than maximizes the usefulness of our [U.S.] sea strength," whereas

American sea power would be "particularly effective in defending Taiwan" itself.[62] Seen from the American perspective in strategic terms alone, even this most militarily useful category of islands was a net liability, especially for an already overextended U.S. Navy.

Category II, comprising two islands in the Dachen group, was not considered important to the defense of Taiwan or the Penghu Islands. Category III's ten smaller islands were mainly important in defending the ten islands in Categories I and II. As for the five other offshore islands under Nationalist domination, U.S. Navy planners noted that they "are not now being utilized for important operations and are not considered worth the effort necessary to defend them against a determined attack." Still, as one U.S. Navy report was quick to point out, none of the offshore islands could be called "essential" to the defense of Taiwan and the Penghus in the sense of being "absolutely necessary" militarily. Their importance to the Nationalists was mainly for "psychological warfare purposes," as well as for their "pre-invasion operations, commando raiding, intelligence gathering, maritime resistance development, sabotage, escape and evasion" (see map 1).[63]

MAP 1
Nationalist-Controlled Offshore Islands

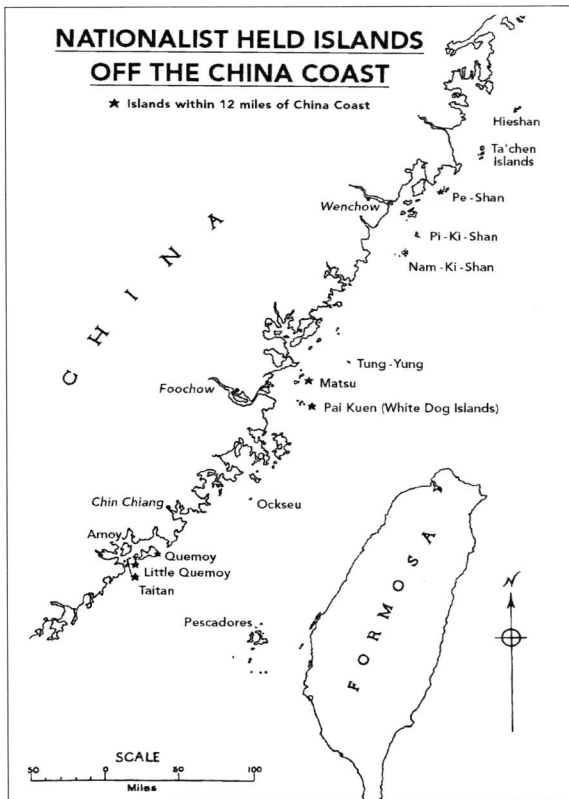

NATIONALIST HELD ISLANDS OFF THE CHINA COAST

★ Islands within 12 miles of China Coast

China's southern coast was especially tense during the early 1950s. Both sides fiercely defended their offshore islands in the hope of changing the strategic balance. The legal status of the disputed islands was unclear. On March 19, 1955, Secretary of State John Foster Dulles argued that since Jinmen and Mazu were indisputably "parts of China," and since the U.S. government recognized the Nationalists as the "Government of China," then the Nationalists had a "better title" to these offshore islands than did the PRC.[64] According to Dulles, "for the United States, the offshore islands were of no intrinsic importance except in the context of an attack on Formosa"; they could be used as "stepping stones for such an attack."[65] Later, the NSC supported this view, suggesting that the number one reason the PRC wanted Quemoy (Jinmen) and Mazu was to remove the Nationalists' "stepping stones toward the mainland."[66]

By the spring of 1950, then, the two Chinas faced each other across the Taiwan Strait. To many it appeared that the PLA was planning to replicate its successful invasion of Hainan Island with a massive maritime attack on Taiwan. The Truman administration's disillusionment with Chiang Kai-shek and his exiled Nationalist government made it highly unlikely that the United States would intervene openly on the side of Taiwan. On October 6, 1949, the NSC reported that while the Nationalists had sufficient supplies to "hold out on Formosa for at least two years," the biggest problem was the "transfer to the Island of the ills and malpractices that have characterized the Kuomintang [the Nationalist political party, led by Chiang since 1926] in China."[67] For his part, General Marshall was convinced "that Chiang's government is both hopelessly corrupt and Fascist in character."[68] Truman's secretary of state, Acheson, described the Nationalist government as "corrupt and incompetent."[69] On October 26, 1949, Truman wrote to Senator Elbert D. Thomas of Chiang Kai-shek's final failure: "Corruption and inefficiency caused that blow up."[70] Five months later Truman told Senator Arthur H. Vandenberg that the "unfortunate situation in the Far East . . . came about as a result of the corrupt Chinese Nationalist government."[71] Trying to defend the weak Nationalist regime, therefore, would be too costly.

Others disagreed. On November 25, 1945, Col. C. A. Seoane told Brig. Gen. Harry H. Vaughan: "Russia's plans now clearly unfolding are to arm and organize this communist group [in China] with a view of having them come into possession of China proper in order that communism shall become definitely established there with its spreading thru Korea and even Japan, thus ousting us eventually from the Pacific, both as to position as well as influence."[72] On January 4, 1950, Hollington K. Tong warned Truman that "once the Communists get into Formosa, there would be no way to prevent Soviet Russia from basing its fleet and submarines in Formosan naval bases—among the best in the world." Russian agents could then "stand at the gates of Japan and the Philippines."[73] Finally, Senator H. Alexander Smith told Truman on May 5, 1950, "My trip to the Far East last fall made it very clear to me that what has happened in China has been the conquest by Russia of that unfortunate country through the Russian method of infiltration and boring from within."[74]

Others had more-nuanced views. H. T. Goodier, a retired American consul who had served in Japan, wrote on January 2, 1950, to the White House that on one hand the U.S. government should not "give any kind of *aid* to Chiang, political, economic or otherwise" but on the other it should "occupy Formosa, pending the conclusion of a peace treaty with Japan and the formation of an autonomous Formosan Chinese state."[75] Many ethnic Taiwanese wanted an independent Formosa, with one group telling Truman:

> Formosa is neither the personal property of Chiang Kai-shek nor the colony of the Communists and inhabitants of Formosa should have a chance and right to determine their own destiny and form of govt; hope this [American] govt will support their demands for the U.N. to send its police forces to take up jurisdiction of Formosa until plebiscite is held for decision as to status of Formosa, and that Formosa remain a permanent, neutral, independent nation under the security of the U.N.[76]

All of this changed in June 1950, when the Korean War began and the political importance of Taiwan suddenly increased.

Conclusions

On January 5, 1950, President Harry S. Truman issued a press statement declaring that the "United States had no predatory designs on Formosa or on any other Chinese territory," and that it would "not provide military aid or advice to Chinese forces on Formosa."[77] Meanwhile, the United States and its allies had "accepted the exercise of [Nationalist] Chinese authority over the island" of Taiwan, did not want to establish a base on Taiwan, and did not have "any intention of utilizing its armed forces to interfere in the present situation." Specifically, it would not become involved in the "civil conflict in China" and would not provide "military aid or advice to Chinese forces on Formosa (Taiwan)." Secretary of State Acheson made clear later that same day that "in the unlikely and unhappy event that our forces might be attacked in the Far East, the United States must be completely free to make whatever action in whatever area is necessary for its own security."[78]

Nonetheless, according to one firsthand report "American personnel"—most likely not directly associated with the U.S. Army—were helping the Nationalists repair and recondition their American-made equipment.[79] Between 1937 and November 1, 1949, the U.S. government had given China $3,523,000,000: $2,422,000,000 in grants, $1,101,000,000 in credits.[80] Much of this aid had been wasted, a matter of "giving aid without proper supervision."[81] But some of this American money had gone for weapons and ammunition, so the Nationalists on Taiwan had military supplies.

Had the PLA in 1950 initiated a spring or summer cross-strait attack, Taiwan might have been incorporated into the PRC. But thanks to their naval dominance in the Taiwan Strait, the Nationalists remained able to carry out an offensive policy, to the extent of capturing cargoes destined for Chinese ports, using the offshore islands to mount raids against the mainland, and to collect valuable intelligence. On May 23, 1950, Sun Li-jen, commander in chief of the Chinese (i.e., Nationalist) army, urged to President

Truman that "Taiwan must be held against the Communists at all costs [as] a beacon light to the suffering people on the mainland of China, who are being trampled upon by the Communists."[82] The Korean War broke out the next month, and suddenly, Taiwan's continued existence was both politically and militarily important. The war could always spread south from Korea and produce a naval invasion of Taiwan. If the PRC did invade Taiwan, it could cut off a major sea line of communications by which United Nations troops and supplies were brought to the Korean theater. That two-prong military threat had to be avoided at all costs.

Notes

1. Claire Lee Chennault [Maj. Gen.] to John R. Steelman, June 10, 1949, Harry S. Truman Papers [hereafter PHST], Official File, OF 150, box 759, File O.F. 150 Misc. (1947–48) [2 of 2], Harry S. Truman Presidential Library [hereafter HSTPL].

2. Senator Homer Ferguson to Harry S. Truman, December 13, 1949, PHST, Official File, OF 150-G, box 761, Formosa, HSTPL.

3. Beginning in 1661, anti-Manchu forces used the various offshore islands to attack and take Taiwan from the Dutch. Two decades later, in 1683, Qing forces used the offshore islands to defeat the Ming loyalists and retake Taiwan. Yang Zhiben [杨志本], ed., *China Navy Encyclopedia* [中国海军百科全书] (Beijing: Sea Tide Press [海潮出版社], 1998), vol. 2, pp. 1912–14.

4. Memorandum for the President, July 26, 1946, Top Secret, PHST, SMOF–Naval Aide, box 22, State Department Briefs File, HSTPL.

5. Ellis E. Patterson to Harry S. Truman, November 14, 1945, PHST, Official File, OF 148-A, box 757, File O.F. 150 (1945–46), HSTPL.

6. Harry S. Truman to Henry A. Wallace, December 18, 1945, PHST, Official File, OF 148-A, box 757, File O.F. 150 (1945–46), HSTPL.

7. Maurice Votaw to Charles G. Ross, April 9, 1947, PHST, Official File, OF 150, box 759, File O.F. 150 Misc. (1947–48) [1 of 2], HSTPL.

8. Summary of Telegrams, June 2, 1947, Top Secret, PHST, SMOF–Naval Aide, box 22, State Department Briefs File, HSTPL.

9. See Bruce A. Elleman, "Huai-Hai," in *The Seventy Great Battles of All Time*, ed. Jeremy

Black (London: Thames & Hudson, 2005), pp. 279–81.

10. Memorandum for the President, November 4, 1948, Top Secret, PHST, President's Secretary's File, box 189, P.S.F. Subject File, HSTPL.

11. Summary of Telegrams, November 8, 1948, Top Secret, PHST, SMOF–Naval Aide, box 23, State Department Briefs File, HSTPL.

12. Summary of Telegrams, November 9, 1948, Top Secret, PHST, SMOF–Naval Aide, box 23, State Department Briefs File, HSTPL.

13. Boston Mayor James M. Curley to Harry S. Truman, November 29, 1948, PHST, Official File, OF 150, box 759, File O.F. 150 Misc. (1947–48) [2 of 2], HSTPL.

14. Frederick C. McKee to Harry S. Truman, November 29, 1949, PHST, Official File, OF 150, box 759, File O.F. 150 Misc. (1947–48) [1 of 2], HSTPL.

15. Members of the National Assembly of the Republic of China to Harry S. Truman, January 6, 1949, PHST, Official File, OF 150, box 758, File O.F. 150 (1947–49), HSTPL.

16. William C. Bullitt to Harry S. Truman, February 9, 1948, PHST, Official File, OF 150, box 759, File O.F. 150 Misc. (1947–48) [1 of 2], HSTPL.

17. Ibid.

18. Chennault to Steelman, June 10, 1949.

19. Central Intelligence Agency, Prospects for a Negotiated Peace in China, August 3, 1948, Secret, PHST, President's Secretary's File, box 216, P.S.F. Intelligence Reports, ORE 12-48, HSTPL.

20. Central Intelligence Agency, Possible Developments in China, November 19, 1948, Secret, PHST, President's Secretary's File, box 216, P.S.F. Intelligence File, ORE 27-48, HSTPL.

21. Summary of Telegrams, December 23, 1948, Top Secret, PHST, SMOF–Naval Aide, box 23, State Department Briefs File, HSTPL.

22. Irving Zuckerman to Clark Clifford, November 30, 1948, PHST, Official File, OF 150, box 759, File O.F. 150 Misc. (1947–48) [2 of 2], HSTPL [emphasis original].

23. Summary of Telegrams, November 26, 1948, Top Secret, PHST, SMOF–Naval Aide, box 23, State Department Briefs File, HSTPL.

24. Central Intelligence Agency, Limitations of South China as an Anti-Communist Base, June 4, 1948, Secret, PHST, President's Secretary's File, box 216, P.S.F. Intelligence File, ORE 30-48, HSTPL.

25. S. C. M. Paine, *The Wars for Asia, 1911–1949* (New York: Cambridge Univ. Press, 2014), p. 269.

26. Summary of Telegrams, January 31, 1949, Top Secret, PHST, SMOF–Naval Aide, box 23, State Department Briefs File, HSTPL.

27. Secretary of State Dean Acheson, Speech to the 35th National Security Council Meeting, March 3, 1949, Secret, PHST, President's Secretary's File, box 179, P.S.F. Subject File, HSTPL.

28. See Bruce A. Elleman, "The *Chongqing* Mutiny and the Chinese Civil War, 1949," in *Naval Mutinies of the Twentieth Century: An International Perspective,* ed. Christopher M. Bell and Bruce A. Elleman (London: Frank Cass, 2003), pp. 232–45.

29. Central Intelligence Agency, Survival Potential of Residual Non-Communist Regimes in China, October 19, 1949, Secret, PHST, President's Secretary's File, box 218, P.S.F. Intelligence File, ORE 76-49, HSTPL.

30. See Bruce A. Elleman and S. C. M. Paine, *Modern China: Continuity and Change 1644 to the Present* (Upper Saddle River, NJ: Prentice Hall, 2010), pp. 343–44.

31. Supplemental Notes on Executive Session, Senate Foreign Relations Committee, October 12, 1949, PHST, President's Secretary's File, box 140, P.S.F. Subject File, HSTPL.

32. Memorandum for the President, December 30, 1949, Top Secret, PHST, President's Secretary's File, box 188, P.S.F. Subject File, HSTPL.

33. While December 7, 1941, is remembered in the United States, in Japan, across the international date line, it was December 8. This timing perhaps emphasized to Western audiences that the Nationalist struggle against communism was a continuation of the Nationalists' alliance with the United States during the Pacific War. Tacit acknowledgment of this may have occurred a year later, on December 8, 1950, when the United States imposed a full strategic embargo on the PRC.

34. Statement by the President, December 18, 1946, PHST, Official File, OF 148-A, box 757, File O.F. 150 (1945–46), HSTPL.

35. Summary of United States Government Economic and Military Aid Authorized for China since 1937, n.d., PHST, Official File, OF 150, box 758, File O.F. 150 (1950–53) [1 of 2], HSTPL.

36. "Intelligence Briefs: China," *ONI Review* (January 1949), p. 39.

37. Central Intelligence Agency, Survival Potential of Residual Non-Communist Regimes in China.

38. See Bruce A. Elleman, "The Nationalists' Blockade of the PRC, 1949–58," in *Naval Blockades and Seapower: Strategies and Counter-strategies, 1805–2005,* ed. Bruce A. Elleman and S. C. M. Paine (London: Routledge, 2006), pp. 133–44.

39. "Appreciation of the Ability of the Chinese Nationalist Navy to Effect a Blockade of Communist Territorial Waters (Secret)," Intelligence Division, Naval Staff, Admiralty, July 9, 1949, FO 371 75902, The National Archives, Kew [hereafter TNA].

40. Central Intelligence Agency, The Ryukyu Islands and Their Significance, August 6, 1948, Secret, PHST, President's Secretary's File, box 216, P.S.F. Intelligence File, ORE 24-48, HSTPL.

41. Central Intelligence Agency, Probable Developments in Taiwan, March 14, 1949, Secret, PHST, President's Secretary's File, box 218, P.S.F. Intelligence File, ORE 39-49, HSTPL.

42. National Security Council [hereafter NSC], The Position of the United States with Respect to Asia, Draft for NSC Staff Consideration Only, October 25, 1949, Top Secret, PHST, President's Secretary's File, box 180, P.S.F. Subject File, HSTPL.

43. Summary of Telegrams, November 21, 1949, Top Secret, PHST, SMOF–Naval Aide, box 23, State Department Briefs File, HSTPL.

44. Omar N. Bradley [Gen.], Chairman of the Joint Chiefs of Staff, Memorandum for the Secretary of Defense, July 27, 1950, Top Secret, PHST, President's Secretary's File, box 181, P.S.F. Subject File, HSTPL.

45. Edward C. Spowart to Harry S. Truman, January 5, 1950, PHST, Official File, OF 150-G, box 761, Formosa, HSTPL.

46. National Intelligence Estimate, Chinese Communist Capabilities and Intentions with Respect to Taiwan, April 10, 1951, Secret, PHST, President's Secretary's File, box 215, P.S.F. Intelligence File, HSTPL.

47. He Di, "The Last Campaign to Unify China," in *Chinese Warfighting: The PLA Experience since 1949,* ed. Mark A. Ryan, David M. Finkelstein, and Michael A. McDevitt (Armonk, NY: M. E. Sharpe, 2003), pp. 73–90.

48. Edward J. Marolda, "Hostilities along the China Coast during the Korean War," in *New Interpretations in Naval History,* ed. Robert W. Love Jr. et al. (Annapolis, MD: Naval Institute Press, 2001), p. 352.

49. D. F. Allen, "Report on Visit to Hong Kong, 15–21 February, 1952 (Secret)," ADM 1/23217, TNA.

50. "The Southeast China Coast Today," *ONI Review* (February 1953), pp. 51–60.

51. NSC, U.S. Action to Counter Chinese Communist Aggression, January 15, 1951, Top Secret, PHST, President's Secretary's File, box 182, P.S.F. Subject File, HSTPL.

52. NSC, Statement of the Problem, January 17, 1951, Top Secret, PHST, President's Secretary's File, box 182, P.S.F. Subject File, HSTPL.

53. James F. Byrnes to Harry S. Truman, February 26, 1946, PHST, Official File, OF 148-A, box 757, File O.F. 150 (1945–46), HSTPL.

54. Harry S. Truman, Executive Order, July 16, 1946, PHST, Official File, OF 150, box 758, File O.F. 150 (1947–49), HSTPL.

55. NSC, Supplementary Measures with Respect to Formosa, March 1, 1949, Top Secret, PHST, President's Secretary's File, box 179, P.S.F. Subject File, HSTPL.

56. William D. Pawley to the Secretary of State, November 7, 1949, PHST, Official File, OF 150, box 758, File O.F. 150 (1947–49), HSTPL.

57. Naval Liaison Office, British Consulate, Tamsui, "American Military Activity in Taiwan (Secret Guard)," October 5, 1951, FO 371/92300, TNA.

58. Frank Holober, *Raiders of the China Coast: CIA Covert Operations during the Korean War* (Annapolis, MD: Naval Institute Press, 1999), pp. 113–14.

59. Robert B. Carney [Adm.], CNO, memorandum to Joint Chiefs of Staff, "Security of the Offshore Islands Presently Held by the Nationalist Government of the Republic of China," July 30, 1953, Top Secret, appendix, Strategic Plans Division, box 289, Naval History and Heritage Command Archives [hereafter NHHC].

60. Preliminary Draft of Possible Statement of Position for Communication to the Republic of China, April 7, 1955, p. 7, Secret, White House Memo [hereafter WHM], box 2, Offshore April–May 1955 (4), Dwight D. Eisenhower Presidential Library [hereafter DDEPL].

61. John Foster Dulles, Draft "Formosa" Paper to Eisenhower, April 8, 1955, p. 9, Secret, WHM, box 2, Offshore April–May 1955 (3), DDEPL.

62. Ibid., p. 7.

63. Carney to Joint Chiefs of Staff, July 30, 1953.

64. John Foster Dulles to Lew Douglas, Personal, March 19, 1955, DDE Ad. Series 12, Douglas, Lewis W. (3), DDEPL.

65. U.K. Embassy, Washington, DC, telegram to Foreign Office (Secret), February 9, 1955, PREM 11/867, TNA.

66. U.S. and Allied Capabilities for Limited Military Operations to 1 July 1961, May 29, 1958, p. C3, Top Secret, WH Office, OSANSA, Records 1952–61, NSC Series, Policy Paper Subseries, box 22, A6 NSC 5724–Gaither Report (1), DDEPL.

67. NSC, The Position of the United States with Respect to Formosa, October 6, 1949, Top Secret, PHST, President's Secretary's File, box 180, P.S.F. Subject File, HSTPL.

68. Senator Glen H. Taylor to Harry S. Truman, March 5, 1948, PHST, Official File, OF 150, box 759, File O.F. 150 Misc. (1947–48) [1 of 2], HSTPL.

69. Secretary of State Dean Acheson, Current Position of the U.S. with Respect to Formosa, August 5, 1949, Top Secret, PHST, President's Secretary's File, box 180, P.S.F. Subject File, HSTPL.

70. Harry S. Truman to Elbert D. Thomas, October 26, 1949, PHST, Official File, OF 150, box 758 (copy in box 826), File O.F. 150 (1947–49), HSTPL.

71. Harry S. Truman to Senator Arthur H. Vandenberg, March 27, 1950, PHST, Official File, OF 150, box 758, File O.F. 150 (1950–53) [2 of 2], HSTPL.

72. C. A. Seoane [Col.] to Harry H. Vaughan [Brig. Gen.], November 25, 1945, PHST, Official File, OF 150, box 758, File O.F. 150 Misc. (1945–46) [1 of 2], HSTPL.

73. Hollington K. Tong to Harry S. Truman, January 4, 1950, PHST, Official File, OF 150, box 759, File O.F. 150 Misc. (1947–48) [1 of 2], HSTPL.

74. Senator H. Alexander Smith to Harry S. Truman, May 5, 1950, PHST, Official File, OF 150, box 758, File O.F. 150 (1950–53) [1 of 2], HSTPL.

75. H. T. Goodier to Harry S. Truman, January 2, 1950, PHST, Official File, OF 150-G, box 761, Formosa, HSTPL [emphasis original].

76. Thomas W. Liao, President, and Frank S. Sins, Chief of Political Affairs, Formosan Democratic Independence Party, to Harry S. Truman, August 9, 1950, PHST, Official File, OF 150-G, box 761, Formosa, HSTPL.

77. Harry S. Truman, statement, January 5, 1950, PHST, Official File, OF 150, box 758 (copy in box 761), File O.F. 150 (1947–49), HSTPL.

78. "Commitments and Problems of the United States to the Republic of China, Including Any Divergencies between the Two Governments, and Chinese Problems with Other Asian Nations (Other Than Military)," n.d. [probably summer 1954], pp. 2–3, Top Secret, WH Office, OSANSA, Special Assistant Series, President Subseries, box 2, Presidential Papers, 1954 (13), DDEPL.

79. Spowart to Truman, January 5, 1950.

80. Summary of United States Government Economic and Military Aid Authorized for China since 1937.

81. Chennault to Steelman, June 10, 1949.

82. Sun Li-jen, Commander in Chief, Chinese Army, to Harry S. Truman, May 23, 1950, PHST, Official File, OF 150, box 758 (summary copy in box 826), O.F. 150 (1950–53) [1 of 2], HSTPL.

CHAPTER TWO

President Harry S. Truman's Decision to Protect Taiwan

[Once] a guard closes a gate, [even] ten thousand [enemy] soldiers cannot open it. (一夫當關, 萬夫莫開.)

With the beginning of the Korean War in June 1950, the U.S. Navy was tasked to create the Taiwan Strait Patrol to neutralize the strait and protect Taiwan from a PRC attack. On August 27, 1950, President Truman emphasized to UN Ambassador Warren R. Austin that "the action of the United States was an impartial neutralizing action addressed both to the forces on Formosa and to those on the mainland. It was an action designed to keep the peace and was, therefore, in full accord with the spirit of the Charter of the United Nations." As President Truman solemnly declared, "We have no designs on Formosa, and our action was not inspired by any desire to acquire a special position for the United States." Furthermore, as a result of this American policy "Formosa is now at peace and will remain so until someone resorts to force."[1]

On September 21, 1950, Secretary of State Acheson explained to the secretary-general of the United Nations the U.S. government's policy toward Taiwan and certain offshore islands, such as the Penghus. He began by citing the statement of the December 1, 1943, Cairo Declaration that Formosa would be returned to "the Republic of China."[2] The July 1945 Potsdam Declaration confirmed Cairo's terms, and the Joint Chiefs' General Order No. 1 directed the Japanese forces on Formosa to surrender to "Generalissimo Chiang Kai-shek." From 1945 onward, the Republic of China retained administrative control over Formosa and the Penghus, plus a number of other offshore islands. On April 28, 1952, Japan signed a peace treaty renouncing its claim to Formosa and the Penghus, although "in whose favor Japan was renouncing was not stipulated." Thereafter, Japan and the ROC signed their own peace treaty, effective August 5, 1952, wherein Japan repeated this renunciation of rights, again without stating to whom.[3]

The ROC had occupied both Formosa and the Penghus continuously since 1945: "There is no question that the Republic of China is today and at all times since the effective date of the Japanese Peace Treaty, has been in effective possession of and exercising administrative control over Formosa and the Pescadores."[4] Nevertheless, its formal title was vague and, as Dulles cautioned Lew Douglas, not 100 percent binding: "I am not quite clear as to why, from a legal standpoint, we are on strong ground as regards Formosa, where the title of the Republic of China is incomplete, than in relation to Quemoy and Matsu where the legal title is complete." Counterintuitively, Dulles argued that if the PRC were to be recognized as the "lawful government of China," then "the legal position of the Chinese Nationalists becomes better in relation to Formosa than in relation to Quemoy and Matsu."[5] As for the PRC taking the islands by force, Dulles told the UN General Assembly: "Also I do not ignore the fact that the Offshore Islands are physically close to Mainland China. But we can scarcely accept the view that nations are entitled to seize territory by force just because it is near at hand." An attempt to take territory that "has long been under the authority of another government . . . is a use of force which endangers world order."[6] This issue became particularly important once the Korean War expanded to include China.

The Korean War

Immediately after the outbreak of the Korean War, President Truman decided that Communist occupation of Taiwan would directly threaten the security of the entire Pacific region. On June 27, 1950, Truman ordered the Seventh Fleet, on the one hand, to intervene to prevent any attack on Taiwan and, on the other, to ensure that the Nationalists "cease all air and sea operations against the mainland." On July 19 Truman told Congress, "The present military neutralization of Formosa is without prejudice to political questions affecting that island. Our desire is that Formosa not become embroiled in hostilities disturbing to the peace of the Pacific and that all questions affecting Formosa be settled by peaceful means as envisaged in the Charter of the United Nations."[7] Truman's attitude toward China and Taiwan was called, accordingly, the "neutralization policy."

The Taiwanese government wanted to solidify relations with the United States and now apparently saw an opportunity. One CIA report had concluded that the Nationalists would use Taiwan as a "bargaining point": "Aware of US interest in that island, they will present themselves as a means and perhaps the sole means of preventing its communization, and will offer various inducements and assurances in return for US aid and US moral support for a regional Chinese regime. They will also argue the legality of such a Chinese administration despite the fact that Taiwan's status has not been formalized [in 1949] by conclusion of a peace treaty with Japan."[8]

Taipei had strong opinions on the neutralization policy. While there were no secret agreements per se, the Nationalist government made its own cooperation with it contingent on four conditions:

> First, pending the conclusion of the treaty of peace on Japan, the Government of the United States of America may share with the Government of the Republic of China the responsibility for the defense of Taiwan.

> Secondly, that Taiwan is a part of the territory of China is generally acknowledged by all concerned Powers. The proposals of the United States of America as contained in the above mentioned *aide memoire* should in no way alter the status of Formosa as envisaged in the Cairo Declaration, nor should it in any way affect China's authority over Formosa.

> Thirdly, the aforesaid proposals and the policies outlined in President Truman's statement dated June 27 are emergency measures adopted to cope with the critical situation as existing on the mainland and in the Pacific region where a number of states have been threatened by or become the victims of aggressive International Communism. The Chinese Government may succeed in the suppression of the aggression of International Communism within a reasonably short time, but should these measures prove to be inadequate, the Chinese Government, in conjunction with other governments concerned, will have to seek for more effective measures in resisting such aggression.

> Fourthly, in accepting the American proposals, the Government of the Republic of China does not intend to depart from its dual policy of resistance against the aggression of International Communism and the maintenance of the territorial integrity of China.[9]

The U.S. Navy assigned a number of ships to form the Taiwan Patrol Force. At any one time, one or more warships patrolled the strait to ensure that the PRC did not attempt to invade. From 1950 through early 1953, the Truman administration ordered the Taiwan Patrol Force to stop attacks from either side of the strait; it was intended to play a neutral role and act as a buffer between the PRC and Taiwan. The neutralization order, however, specifically did not include the many offshore islands controlled by the Nationalists. On October 7, 1950, it was made clear to the forces involved that they would not participate in the "defense of any coastal islands held by the Nationalist Chinese nor will they interfere with Nationalist Chinese operations from the coastal islands."[10]

Shortly after Dwight D. Eisenhower won the 1952 presidential election, the focus of the Taiwan Patrol Force began to change. Eisenhower heeded hard-liners' calls to "unleash Chiang Kai-shek."[11] A nationwide Gallup poll early in 1953 showed that 61 percent of Americans supported "the United States supplying more warships to Free China for use in blockading the China mainland coast and more airplanes for use in bombing the China mainland."[12] On February 2, 1953, Eisenhower announced in his State of the Union address that there is "no longer any logic or sense in a condition that required the United States Navy to assume defensive responsibilities on behalf of the Chinese Communists." As for the offshore islands, John Moore Allison, Assistant Secretary of State for Far Eastern Affairs, explained that they were not included in the original order "or in the contemplated amendment of the order."[13]

This policy change opened a new peripheral theater in the south, meant to pressure Beijing to sign a peace treaty ending the Korean War.[14] As early as April 5, 1951, the Joint Chiefs of Staff had concluded that "preparations should be made immediately for action by naval and air forces against the mainland of China."[15] On July 31 Senator Harry P. Cain had urged Truman to support Nationalist attacks in the south as a "potentially powerful means of strengthening our own hand in the truce negotiations and [one that] may, if adequately supported, force some Communist withdrawals from the Korean front."[16] In 1954, a top-secret State Department study acknowledged that this new policy "represented the then need for diversionary threats."[17] Dulles would revert to this diversionary strategy, suggesting "that it might be preferable to slow up the Chinese Communists in Southeast Asia by harassing tactics from Formosa and along the seacoast which would be more readily within our natural facilities than actually fighting in Indochina." The transcript of the conversation records that "the President indicated his concurrence with this general attitude."[18]

After the Chinese communists intervened in the Korean War in October 1950, the Joint Chiefs of Staff recommended to the secretary of defense a stronger Taiwan policy:

   a. Maintain the security of the off-shore defense line: Japan—Ryukyus—Philippines.

   b. Deny Formosa to the Communists.[19]

Three days later, a top-secret NSC report went a step farther: "Send a military training mission and increase [Mutual Defense Assistance Program] to the Chinese on Formosa."[20] By March 22, 1952, the Joint Chiefs of Staff had developed five points in support of Taiwan:

   a. Take such measures as may be necessary to deny Formosa to any Chinese regime aligned with or dominated by the USSR;

   b. In its own interests, take unilateral action if necessary, to insure the continued availability of Formosa as a base for possible United States military operations;

   c. Continue that part of the mission presently assigned to the 7th Fleet relative to the protection of Formosa until such time as conditions in the Far East permit the Chinese Nationalists on Formosa to assume the burden of the defense of that island;

   d. Support a friendly Chinese regime on Formosa, to the end that it will be firmly aligned with the United States; and

   e. Develop and maintain the military potential of that Chinese regime on Formosa.

The primary goal of these five points was "the denial of Formosa to communism," which was "of major importance to United States security interests, and . . . of vital importance to the long-term United States position in the Far East."[21]

On February 2, 1953, Eisenhower lifted the U.S. Navy's previous orders to restrain Nationalist forces. In his State of the Union address, as noted, Eisenhower characterized the neutralization policy as helping the Communists, as telling them, "in effect, that the

United States Navy was required to serve as a defensive arm of Communist China."²²
But the U.S. Seventh Fleet would "no longer be employed as a shield for the mainland
of China."²³ Nevertheless, as the attorney general later made clear, the new order to the
Navy was not "evidence of any understanding that the President's announcement meant
that the United States would defend the off-shore islands controlled by the Nationalists
in addition to defending Formosa and the Pescadores."²⁴

One immediate result of this policy change was that the Commander, Seventh Fleet
was ordered no longer to require the Taiwan Patrol Force to prevent the use of Taiwan
and the Penghu Islands as Nationalist bases for operations against the mainland.²⁵ This
change led to the revision of previous orders to the force itself. In particular, a formal
"pen-and-ink change" to Operation Order 20-52, "Special Patrol Instructions," directed
deletion of "Large forces moving from Formosa toward the mainland will be reported to
[Commander, Task Group] 72.0."²⁶ Most congressmen agreed with this change of policy,
but Senator John Sparkman (D-AL) pointed out that because the previous policy had
never "prevented 'pin-prick' raids from Quemoy or the off-shore islands," the new order
could be seen as a "medium for enlarging the war."²⁷

In April 1953, talks were held in Taipei between Adlai Stevenson and Chiang Kai-shek.
Chiang promised Stevenson that given continued American military support, his forces
would be ready to return to the mainland within "three years at the latest" and that once
there they would gain a significant domestic following within "three to six months."²⁸
The U.S. government agreed to support this plan. One reason later given by Dulles to
the British ambassador was the "peripheral" view, that it would "make a diversionary
threat at a time when fighting was going on in Korea so as to cause the Chinese Communists to transfer forces away from Korea towards Formosa."²⁹

The Nationalists seemed eager to begin. The first raid mounted directly from Taiwan
was against Dongshan Island, which had been taken from the Nationalists on May 11,
1950, by an assault of over ten thousand PLA troops. In mid-July 1953, the Nationalists
tried to retake it with approximately 6,500 guerrillas, marines, and paratroopers. The
attempt failed, but it forced the PRC to relocate troops south. Such results of Eisenhower's decision to open a peripheral campaign played an important role in the PRC's
willingness to open talks in Korea. After the armistice was signed on July 27, 1953, the
PRC immediately began to move additional troops south and station them across from
Taiwan. This would lead to escalation on both sides of the Taiwan Strait.

America's Pacific Allies and Great Britain

Spurred on by war on the Korean Peninsula, the U.S. Navy's patrols "neutralized" the
Taiwan Strait. They also initially discouraged the Nationalists from mounting a major
attack on the Chinese mainland. The goal, as noted, was to limit the possibility of the

Korean conflict spreading farther to the south and possible escalation into a world war between the United States and the USSR. Later, also as noted, tensions were increased along the strait as a diversion; as Dulles told a New Zealand delegation, "unleashing Chiang" would "encourage the Chinese [Communists] to retain substantial forces opposite Formosa."[30]

Great Britain opposed this policy, but America's Pacific allies generally supported it. The U.S. government had to be concerned about its Pacific allies. As Dulles warned in March 1955, "If Formosa were lost, that would almost surely involve such a major breach of the offshore island chain that the whole position would be lost and the Pacific defense of this continent would be forced back close to our own Pacific Shores." Should that happen, the "situation in Australia and New Zealand would become virtually untenable."[31] Later that month, Eisenhower wrote to Churchill that "the loss of Formosa would doom the Philippines and eventually the remainder of the region," and that in American "contacts with New Zealand and Australia, we have the feeling that we encounter a concern no less acute than ours."[32] For all of these reasons, keeping Taiwan out of communist hands was considered vital.[33]

Australia and New Zealand were usually counted as parts of the anticommunist defense line in the Pacific, from the Aleutians to "Japan–Ryukyus–Philippines–Australia and New Zealand."[34] In 1955 the Australian prime minister, Robert Menzies, put it succinctly: "From the point of view of Australia and, indeed, Malaya, it would be fatal to have an enemy installed in the island chain so that by a process of island hopping Indonesia might be reached and Malaya and Australia to that extent exposed to serious damage either in the rear or on the flank."[35] A year later the Japanese foreign minister, Mamoru Shigemitsu, told the American ambassador that "Japan would consider the fall of Taiwan to the Communists as a threat to its interests and therefore supports the U.S. policy of preventing such an eventuality."[36]

The United Kingdom was more difficult. The United States and Britain were fighting side by side in Korea, but the Americans were selectively embargoing the PRC (as discussed in detail below) while the British were trading with it. In early 1951, largely in response to China's armed intervention in the Korean War but also with due consideration for the Nationalist blockade, London considered limiting trade with the PRC but decided that British workers would "undoubtedly suffer severely from any stoppage and would have difficulty in finding alternative employment." Not only could an embargo lead to the permanent loss of shipping companies' holdings in the PRC, but British ships in Chinese ports, if not given sufficient advanced warning (a minimum of five days, preferably ten), might be trapped there and at risk of seizure. At any one time there could be as much as 120,000 tons of British-owned shipping in Chinese ports, a factor that "reinforces the absolute need for notice if British tonnage is to be saved from falling into Chinese hands."[37] Another consideration for Britain was the long-term future of its trade with China. A complete embargo of China would also devastate Hong Kong, since approximately 45 percent of all of Hong Kong's exports went to China. According to the governor of Hong Kong, without this trade the "thriving, trading, financial and insurance entrepôt of the Colony" would quickly become "an economic desert."[38]

Considering all this and their large share in the China market, British shippers felt compelled to maintain trading. But that meant British ships had to challenge the Nationalist navy's blockade. The Nationalists staged attacks from offshore islands against the mainland ports, Communist-held islands, and convoys escorted by PRC junks armed with small artillery pieces, mortars, and automatic weapons. Whenever possible, Nationalist ships would attack and, when they could, surround and sink the convoys. The economic impact (aside from that of the U.S. embargo on strategic goods) was significant: the blockade intercepted a high percentage of China's international trade. Between 1950 and 1952 the Nationalists halted and searched some ninety ships heading for Communist ports, two-thirds of them British-flagged ships registered in Hong Kong. The British government protested vehemently the seizure of cargo, arguing that its ships were complying with American limits on strategic goods and so were carrying "only nonstrategic cargoes."[39]

During the late 1940s and early 1950s the United States and the United Kingdom discussed almost continuously what to do about China and Taiwan. On February 10 and March 22, 1949, American and British officials met regarding controls on foreign exports to China.[40] On August 1, 1949, the U.S. embassy in London was instructed to tell the British government that "we feel that the failure to demonstrate effective western control over exports of key importance to China's economy would represent abandonment of the most important single instrument available for the defense of western interests in the Far East." The embassy was to urge London to back restrictions on both restricted and prohibited lists (referred to below as "1A" and "1B"), "since the mutual security interests of the west must be considered as much in terms of political and economic strategy as in terms of direct military factors."[41]

In November 1949, a report commissioned by Secretary of State Acheson stated, "We have been reluctant thus far to impose unilaterally new controls over exports to China because of the possibility that such action would handicap our negotiations with the British." After it had been pointed out to them "that such controls would represent the most important single instrument available for use vis-à-vis the Chinese Communists," the British had finally agreed to control "1A list exports to China," provided that the other major powers also did so; to limit petroleum sales; and to "observe and to exchange information with the United States . . . on the movement of 1B goods to China with a view to joint consultation regarding corrective measures if it appeared that the flow was excessive or injurious to our common interests."[42]

To avoid undermining Anglo-American relations, it was eventually decided mutually to leave the embargo of strategic goods largely to the United States, with help from the Nationalist navy. With tacit American support, the Nationalist forces tightened the blockade throughout 1949 and into 1950. After the outbreak of the Korean War and particularly the PRC intervention, the U.S. government considered a total naval blockade of its own against mainland China. Of course, the British were concerned about Hong Kong; at any point, the PRC could take it by force. Hong Kong's vulnerability led the British to recognize the PRC on January 6, 1950, and to adopt a more liberal trade policy with it. But by 1958 the British foreign secretary, then Selwyn Lloyd, would reaffirm that the United Kingdom and United States shared the view that there was a "Communist menace in the Far East," and that the "containing line" had to include Japan, South Korea, Okinawa, Taiwan, Hong Kong, South Vietnam, and Malaya.[43]

American-Led Strategic Embargo of the PRC

A full naval blockade of the PRC was politically infeasible, owing to the possible retaliation of the USSR, on the one hand, and to Hong Kong's sensitive strategic position, on the other. Beginning in December 1950, the U.S. government instead began to impose an embargo on strategic goods to the PRC. On December 16, 1950, it announced the issuance of "regulations to prohibit ships of United States registry from calling at Chinese Communist ports until further notice," as part of an ongoing effort to halt all exports "to Communist China from the United States without validated export licenses."[44] President Truman signed an executive order implementing the Export Control Act of 1949, which had in effect reoriented 1940 restrictions aimed at Japan toward the Soviet bloc. This policy was subsumed under the Coordinating Committee for Multilateral Export Controls (COCOM), a trade-control regime adopted in January 1950.

Americans largely supported this embargo. On December 1, 1950, a telegram from an ordinary citizen in San Francisco arrived at the White House addressed to Maj. Gen. Harry H. Vaughan:

> Thirty-one cases automobile ignition parts on pier thirty-one in San Francisco consigned to Communist China will move by Pacific Transport Lines ship China Bear about December 3. Handled by Hawaiian freight forwarders. This shipment is valuable as war material. Customs authorities claim they cannot stop. As a former member of 129 field artillery and supporter of President Truman I urgently request Presidential Order to hold this shipment or do you suggest another Boston tea party.

Six days later, Major General Vaughan wrote to thank the sender and to report, "Immediately upon receipt, we contacted the Commerce Department; and today we have been advised that they were successful in stopping the shipment of the parts you described in your telegram."[45]

On May 18, 1951, the United Nations adopted a selective embargo prohibiting the sale of "arms, ammunition and implements of war, atomic energy materials, petroleum, transportation materials of strategic value, and items useful in production of arms, ammunition and implements of war." With that action, in conjunction with all the other unilateral and international restrictions, there had been "instituted and maintained controls on strategic trade with Communist China that are much more severe and sweeping than the system applicable to Soviet Russia and the rest of the Soviet bloc."[46] Although the strategic embargo was less effective than a full naval blockade would be, the U.S. government worked closely with the Nationalists to enforce it.

The fifteen-country COCOM group was composed of the United States, all the NATO countries (minus Iceland), and then also Japan. To convince other countries to conform to these proscriptions against China, Congress adopted the Mutual Defense Assistance Control Act of 1951. Commonly called the "Battle Act," after its sponsor Congressman Laurie C. Battle, this legislation would terminate economic and military aid to "countries which fail to cooperate in the control program."[47] After China intervened in the Korean War, the restrictions became even tougher. During the fall of 1952, the China Committee (CHINCOM) instituted even tighter controls, embargoing industrial machinery, steel-mill products, and metal of all types. Meanwhile, the U.S. embargo was taken up by more countries in COCOM and CHINCOM, such as Greece and Turkey in 1953, and bolstered by pledges of "cooperation from important neutral countries, notably Sweden and Switzerland."[48] The United States shouldered the lion's share of the financial burden of enforcing sanctions, however, including an estimated $1.7 billion every year to various East Asian countries in outright aid "designed to secure our own vital defense arc in the Far East stretching from the Aleutians to Southeast Asia."[49]

On January 12, 1951, the Joint Chiefs of Staff agreed to "continue and intensify now an economic blockade of trade with China."[50] They also advised the secretary of defense "that the neutralization of Formosa would not meet United States military strategic needs since it would improve the strategic position of the Communists by releasing their forces for build-up, and would reduce the strategic position of the United States and restrict freedom of action in the event the military situation required an armed attack against the Chinese Communists on the Mainland."[51] Gen. Walter B. Smith, Director of Central Intelligence, told Truman on September 24, 1952, that "building up a Chinese Nationalist force on Formosa was perfectly safe because they will be unable to go anywhere unless we release the trigger in the form of air and naval support."[52]

An early study of the sanctions program pointed out that what the USSR wanted from the West were "items having a high skilled labor content." While it would be too costly to cut all trade in such goods to Russia, the situation in China was different, because "China's trade is so vulnerable to the disruption of its sea communications." That in turn meant, "Because of the nature of China and the advance state of military activities in that area, our program of economic warfare should be more extensive than in Europe [and] we should be less reluctant to take necessary all-out measures. An important objective would be the attainment of multilateral agreement among the Western Nations on a complete quarantine of China from the technology of the West." In devising such a quarantine, "careful consideration should be given to an informal merchant shipping embargo of Chinese ports."[53]

There was an important second side of this coin: even while limiting the Soviet bloc countries' imports, the United States needed to assist the economic growth of Western-aligned countries. On April 26, 1951, new language was added to the NSC 48/3 report. Below point 5(d), on helping East Asian countries with their security concerns and economic relations, a new subparagraph appeared: "h. In accordance with 5(d) above, take such current and continuing action as may be practicable to maximize the availability of the material resources of the Asian area to the United States and the free world generally, and thereby correspondingly deny these resources to the communist world."[54]

A CIA report later assessed that free-world "controls have somewhat retarded the development of China's economic potential." After discussing the possibility of a full naval blockade to cut China's sea lines of communications or air attacks on crucial inland lines, this report warned that the former might result in conflict with the USSR over its access to Port Arthur and Dalian. There was also a chance that China might retaliate, by "seizing Hong Kong and Macao."[55] Instead of either plan, the strategic embargo had to be tightened by broadening its reach. One report even suggested: "Encourage the countries of Southeast Asia to restore and expand their commerce with each other and with the rest of the free world, and stimulate the flow of the raw material resources of the area to the free world."[56]

Although, as a confidential 1955 U.S. Navy report admitted, the strategic embargo was "not complete" and "China [was] actually obtain[ing] differential goods through triangular deals and transshipments," the overall success of the embargo was evident in the facts that "Chinese procurement has been hampered and the total amount of goods purchased is reduced as a result of higher costs."[57] A 1955 report to Eisenhower persuasively argued for keeping "the Chinese Communist regime under economic (and other) pressures. . . . Such pressures add to the strains which can ultimately lead to disintegration." This was particularly true with regard to the PRC's foreign commitments to its allies, since failure to grow at a rapid pace would mean Beijing could not fund these commitments: "This kind of dilemma tends to lead to a breakdown."[58]

Impact of the American-Led Embargo

The embargo could be effective only if Washington convinced most of the other Western trading nations to join it. By and large, in fact, America's allies backed the strategic embargo, at least in part. For example, the British China Association reported to the Foreign Office in February 1951 that the American sanctions program was quite effective and had produced a "great shock to [the mainland] Chinese economy."[59] The Swedish ambassador to China likewise confirmed during June 1951 that the strategic embargo was having the desired effect, since international "shipping was the Achilles' heel of China," on which "a very effective squeeze could be placed."[60]

Nevertheless, estimating the effectiveness of the strategic embargo was difficult. How were the various costs the program imposed on China, some of them—such as political and psychological stress—not measurable in dollars, to be calculated? In late 1956, an estimate put the costs to China of the U.S. part of the embargo at $100–$150 million per year in trade losses, plus $100 million in frozen financial assets. With regard to the other CHINCOM members, the estimate was between $115 and $154 million. Together, this meant estimated losses in the $215–$304 million range, which represented a 13–18 percent reduction in China's strategic imports. A large percentage of these added costs, perhaps as much as a third, was due to increased expense of transportation, now mainly via the Trans-Siberian Railway, which stressed the PRC's diplomatic relations with the USSR.[61]

In addition to financial costs, sanctions had other effects that benefited Western interests. For example, the sanction program cut off the PRC from convertible foreign currencies, imposed "serious and costly delays in Communist Chinese economic planning and implementation," and put a heavy strain on China's "existing equipment and facilities," in particular those related to transportation. Finally, that the sanctions on the PRC were above and beyond those on the rest of the communist bloc made China a "pariah" state, which added even more "political, psychological, and prestige 'costs' to Communist China."[62]

The obverse was that reducing completely, or perhaps even partially, the economic and trade sanctions on China would have a number of foreseeable hidden costs for the West. First, it would mean an "enormous loss of prestige" (read: face) to the United States, especially in East Asia. Second, many neutral countries might move closer to the PRC. Third, it might stimulate even greater "Communist subversive activities throughout Asia." Fourth, it might lead to the loss of important U.S. bases in the region. Fifth, it might result in the PRC's admission to the United Nations, perhaps even "displacing Free China on the Security Council." Sixth, it might reduce the chance for "securing the release of Americans imprisoned in Communist China." Seventh, it would mean the "loss without adequate recompense of an important bargaining weapon in the cold war in the Pacific." Eighth, it would entail a "fundamental undermining of our moral position against aggression and oppression." In short: "Removal of the China trade differential would therefore relieve Communist China of a very onerous and costly burden."[63]

The effectiveness of the embargo, however, had a negative impact on British trade with China, which led to a sharp increase in Anglo-American tension. During bilateral talks, the British disagreed with the Americans on what should be considered strategic trade. Accordingly, by late 1956 "the British [had] seriously suggested to us, at responsible levels, that the only 'defensible' criterion for classifying exports to Red China as 'strategic,' is whether or not they could contribute directly to thermo-nuclear warfare." Meanwhile, other CHINCOM partners decried the use of a "stick" alone and sought to offer China "carrots" as well: "There are important government officials and elements of public opinion among them who believe that the closer the Free World countries can get to the Red Chinese, the better the chance of converting them from their [aggressive] designs on Asia."[64]

In October 1952, the British government was considering changes to its own export-control policies. In 1953, as a direct response to the Nationalist blockade, the Royal Navy was ordered to form its own "Formosa Straits Patrol" to protect British commercial ships. There were some contradictory voices. On June 18, 1953, for example, Roger Makins told John Foster Dulles that the British government would agree to maintain economic pressure on the PRC, in particular because it offered "trading ground at a [future] Korean political conference."[65] One major difference, therefore, was over the political objectives of each country vis-à-vis Beijing.

The number of Nationalist attacks gradually decreased following the creation of the Royal Navy's Formosa Straits Patrol. During 1953–54, the Nationalist blockade gradually shifted away from patrol vessels toward airpower, mainly provided by the United States. In April 1953, Taipei also adopted more-stringent shipping regulations, which, largely in line with those already promulgated by the U.S. Maritime Shipping Association, would "prohibit any government-chartered foreign vessel from proceeding to any country behind the Iron Curtain within a 60-day period after it had discharged its cargo at ports in Free China."[66] Chiang Kai-shek wrote to President Eisenhower on April 15, 1953, that it was time to "gradually take away the political and military initiative from the Communist world."[67]

During late July 1954, the PRC shot down a Cathay Pacific Airways airliner. When Eisenhower was told that the British (the airline was based in Hong Kong) were playing it down and accepting Chinese protestations that it had been an accident, he replied, "I wonder if old John Bull will take that."[68] The U.S. Navy sent two aircraft carriers to the scene. Initially, their pilots were told not to invade Chinese airspace, even in "hot pursuit." At Dulles's urging, Eisenhower changed this policy. U.S. Navy ships and planes could now take "all necessary measures to protect themselves," including pursuing an enemy ship or plane into Chinese sovereign waters.[69] On July 26, Dulles informed Eisenhower that two American search planes had been shot down near Hainan; the president's reaction was, "Well, it didn't take long for that to happen." Coordinating with the British government official protests to the PRC was all-important: "If we adopt a stiff one [i.e., posture] and the British not, it will cause more friction between our countries."[70]

Such concerns were well-founded. In September of that year the State Department told Eisenhower that Sam Watson, a member of the British Labour Party delegation to Moscow, was warning that a Chinese attack on Taiwan would have the "single strategical aim of splitting West," specifically to shatter "US–United Kingdom–French cooperation." Furthermore, Zhou Enlai (then both premier and foreign minister of the PRC) had told Watson that liberating Taiwan was his government's "most important problem," equal to if not greater in criticality than "feeding, clothing, and housing their millions" of citizens.[71] Its urgency was reflected in the fact that in 1954 the Nationalist air force attacked British shipping on thirty-two occasions.[72]

Conclusions

In 1951, a full American blockade of the PRC was discussed by U.S. Navy planners, but fear of undermining the Anglo-American alliance both in Europe and in Korea overshadowed any possible benefits. During late 1953 and early 1954, therefore, the U.S. government provided the Nationalists better equipment with which to enforce the blockade from the air. But that upset the British, their colony in Hong Kong being a major reason. But as Eisenhower reminded Churchill, the United States had, on the one hand, no colonies to worry about but, on the other, an anticolonialist position to reinforce: "We have no possessions in that immediate area. Consequently, we cannot be accused of any support of colonialism or of imperialistic designs."[73] This echoed a June 13, 1946, letter from Truman to no less a figure than the American author Pearl Buck, assuring her that after the war in Asia "we have asked for no territory and we have asked for no reparations."[74] The Chinese people were being "deceived into serving the ends of the Russian colonial policy in Asia."[75]

In fact, Eisenhower had tried, and failed, to convince Churchill that "colonialism was ended" and that the British colonies should be given a chance to stay or leave the British

Empire.[76] If conflict over the offshore islands ever broke out, it would be "primarily a fight between the Chinese Nationalists and the Chinese Communists, and not a fight between the 'white' Western and the 'yellow' Chinese" worlds; it would "not take on the appearance of a struggle between races."[77] But Whitehall feared that too strict an embargo might spark that war.[78] Because Great Britain had recognized the PRC, the "British people as a whole look on the offshore islands as belonging to Red China, and consider that we are foolish to be supporting Chiang even indirectly in possession of those areas."[79]

An even greater British concern was that increasing Anglo-American friction "might prejudice world-wide defence cooperation between the United Kingdom and United States, with possibly serious consequences to the security of Western Europe, the United Kingdom and, ultimately, the United States."[80] There was always a chance that China would reject monolithic communism. As long as the PRC could expect to obtain certain supplies from the West, "she would not be completely dependent on the Soviet Union and the East European satellites to meet her needs."[81] Indeed, in 1954–55, during the "first" Taiwan Strait crisis, one of the PRC's most important objectives was to force the Nationalists to end their blockade of the Chinese coast.

Notes

1. Harry S. Truman to UN Ambassador Warren R. Austin, August 27, 1950, PHST, Official File, OF 150-G, box 761, Formosa, HSTPL.

2. Dean Acheson to Dulles, February 1, 1955, Dulles, J. F., WHM, box 2, WHM 1955, FS (2), DDEPL.

3. Ibid.

4. Ibid.

5. Dulles to Douglas, March 19, 1955.

6. Address by the Honorable John Foster Dulles, Secretary of State of the United States of America before the United Nations General Assembly, September 18, 1958, D-H 10, Dulles Sept. 1958 (1), 2, DDEPL.

7. "Commitments and Problems of the United States to the Republic of China," p. 3.

8. Central Intelligence Agency, Probable Developments in Taiwan, June 16, 1949, Secret, PHST, President's Secretary's File, box 218, P.S.F. Intelligence File, ORE 45-49, HSTPL.

9. "Commitments and Problems of the United States to the Republic of China," p. 4, citing

China Handbook 1951 (Taipei: China Publishing, 1951), p. 115.

10. ComCruDivONE Operation Order No. 7-50, October 7, 1950, Post-1946 Operation Plans, Task Force 72, NHHC.

11. "First Taiwan Strait Crisis: Quemoy and Matsu Islands," GlobalSecurity.org, n.d., accessed December 14, 2010, www.globalsecurity.org/military/ops/quemoy_matsu.htm.

12. "U.S. to Send 100 Thunderjets, Warships to Free China," February 4, 1953, Free China Information, FO 371/105272, TNA.

13. "Commitments and Problems of the United States to the Republic of China," p. 8.

14. For more on opening new peripheral theaters, see Bruce A. Elleman and S. C. M. Paine, eds., Naval Power and Expeditionary Warfare: Peripheral Campaigns and New Theatres of Naval Warfare (London: Routledge, 2011).

15. Omar N. Bradley [Gen.], Joint Chiefs of Staff, Military Action in Korea, April 5, 1951, Top Secret, PHST, President's Secretary's File, box 183, P.S.F. Subject File, HSTPL.

16. Senator Harry P. Cain to Harry S. Truman, July 31, 1951, PHST, Official File, OF 150, box 758, File O.F. 150 (1950–53) [1 of 2], HSTPL.

17. Report on U.S. Government Policies in Relation to China, n.d., p. 4, Top Secret, Dulles, J. F., WHM, box 2, 1954, Formosa Straits (1), DDEPL.

18. Memorandum of Conversation with the President, March 24, 1954, Dulles, J. F., WHM, box 1, Meetings 1954 (4), DDEPL.

19. Joint Chiefs of Staff, Courses of Action Relative to Communist China and Korea, January 12, 1951, Top Secret, PHST, President's Secretary's File, box 182, P.S.F. Subject File, HSTPL.

20. NSC, U.S. Action to Counter Chinese Communist Aggression.

21. NSC, Joint Chiefs of Staff, On Formosa, March 24, 1952, Top Secret, PHST, President's Secretary's File, box 186, P.S.F. Subject File, HSTPL.

22. "Memorandum by the Attorney General on Congressional Attitude to Formosa Defense," n.d. [probably summer 1954], WH Office, OSANSA, Special Assistant Series, President Subseries, box 2, Presidential Papers, 1954 (7), DDEPL.

23. U.K. Embassy, Washington, DC, to Foreign Office (Secret), January 30, 1953, PREM 11/867, TNA.

24. "Memorandum by the Attorney General on Congressional Attitude to Formosa Defense."

25. CTF 72 Operation Order No. 2-A-53, May 1, 1954, Post-1946 Operation Plans, Task Force 72, NHHC.

26. CTG 72.0 Operation Order No. 20-52, December 3, 1952, Post-1946 Operation Plans, Task Force 72, NHHC. There is no way to know when this line was marked out, but most likely it was during January–March 1953.

27. "Memorandum by the Attorney General on Congressional Attitude to Formosa Defense."

28. John Foster Dulles Papers [hereafter JFDP], April 8, 1953, reel 204/205, 88851, Princeton Univ.

29. U.K. Embassy, Washington, DC, telegram to Foreign Office (Secret), October 18, 1954, CAB 21/3272, TNA.

30. "Extract of New Zealand Delegation's Record of Discussions of Anzus Council Meeting in Washington, September 1953 (Secret)," FO 371/105272, TNA.

31. Summary of Remarks of the Honorable John Foster Dulles at Cabinet Meeting Ottawa, March 18, 1955, p. 2, Top Secret, DH 5, Dulles, J. F., March 1955, DDEPL.

32. President Eisenhower to Prime Minister Churchill, March 29, 1955, Top Secret–Eyes Only, DDE Diary Series 10, DDE Diary March 1955 (1), 5, DDEPL.

33. [B. A.?] Fraser, P. S. Slessor, and W. J. Slim, "Meeting of Prime Ministers: The Strategic Importance of Formosa; Memorandum by the United Kingdom Chiefs of Staff (Top Secret)," January 6, 1951, PREM 8/1408, TNA.

34. NSC, United States Objectives, Policies and Courses of Action in Asia, May 17, 1951, Top Secret, PHST, President's Secretary's File, box 183, P.S.F. Subject File, HSTPL.

35. "Formosa and Off-Shore Islands, Note by the Prime Minister of Australia (Secret)," Meeting of Commonwealth Prime Ministers, February 8, 1955, PREM 11/867, TNA.

36. JFDP, December 2, 1956, reel 212/213, 94413, Princeton Univ.

37. Effect on British Shipping of Stoppage of Trade with China, January 16, 1951, FO 371/92273, TNA.

38. "Sanctions against China. Probable Economic Political and Strategic Consequences in Hong Kong, Malaya and South East Asia Generally," Draft Memorandum (Top Secret), n.d. [most likely December 1950], FO 371/92276, TNA.

39. "Southeast China Coast Today."

40. Dean Acheson to Sidney W. Souers, Implementation of NSC 41, April 14, 1950, Top Secret, PHST, President's Secretary's File, box 182, P.S.F. Subject File, HSTPL.

41. Summary of Telegrams, August 1, 1949, Top Secret, PHST, SMOF–Naval Aide, box 23, State Department Briefs File, HSTPL.

42. Dean Acheson to Sidney W. Souers, U.S. Policy Regarding Trade with China, November 4, 1949, Top Secret, PHST, President's Secretary's File, box 182, P.S.F. Subject File, HSTPL.

43. Selwyn Lloyd to John Foster Dulles (Top Secret), September 11, 1958, CAB 21/3272, TNA.

44. U.S. State Dept., Control of United States Economic Relationships with Communist China, December 16, 1950, PHST, President's Secretary's File, box 182, P.S.F. Subject File, HSTPL.

45. George O. Kelly, telegram to Harry H. Vaughan, December 1, 1950, and Vaughan to Kelly, December 7, 1950, PHST, Official File, OF 150, box 758, File O.F. 150 (1950–53) [1 of 2], HSTPL.

46. John Foster Dulles to J. Bracken Lee, Governor of Utah, December 30, 1953, D-H-2, Dulles Dec. (53), DDEPL.

47. "Southeast China Coast Today."

48. Sherman R. Abrahamson, "Intelligence for Economic Defense," *Central Intelligence Agency: Library,* www.cia.gov/library/center-for-the -study-of-intelligence/kent-csi/vol8no2/html/ v08i2a03p_0001.htm.

49. The Case for Maintaining a Meaningful China Trade Control Differential, n.d., p. 1, Secret, DDE U.S. Council on For. Econ. Policy, Randall Series, Trips Subseries, box 2, Far East Trip [Dec. 1956], Background Papers (2), DDEPL.

50. Joint Chiefs of Staff, Courses of Action Relative to Communist China and Korea.

51. Joint Chiefs of Staff, China Lobby, General, January 2, 1951, Secret, PHST, President's Secretary's File, box 140, P.S.F. Subject File, HSTPL.

52. Memorandum for the President, September 24, 1952, Top Secret, PHST, President's Secretary's File, box 189, P.S.F. Subject File, HSTPL.

53. Economic Cooperation Administration, Trade of the Free World with the Soviet Bloc, February 1951, Secret, PHST, President's Secretary's File, box 183, P.S.F. Subject File, HSTPL.

54. NSC, United States Objectives, Policies and Courses of Action in Asia, April 26, 1951, Top Secret, PHST, President's Secretary's File, box 183, P.S.F. Subject File, HSTPL.

55. Central Intelligence Agency, Probable Effects of Various Possible Courses of Action with Respect to Communist China, June 5, 1952, Top Secret, PHST, President's Secretary's File, box 186, P.S.F. Subject File, HSTPL.

56. NSC, United States Objectives, Policies and Courses of Action in Asia, June 25, 1952, Top Secret, PHST, President's Secretary's File, box 186, P.S.F. Subject File, HSTPL.

57. W. K. Smedberg, Director, Politico-Military Policy, "Memorandum of Information for the Secretary of the Navy" (Confidential), October 13, 1955, Strategic Plans Division, box 326, NHHC.

58. Maintenance of Economic Pressures against Communist China, n.d., Secret, WH Office, OSANSA, Special Assistant Series, President Subseries, box 2, Presidential Papers 1955 (5), DDEPL.

59. China Association, London, to J. S. H. Shattock, Foreign Office, London, February 26, 1951, FO 371/92276, TNA.

60. Summary of Telegrams, June 5, 1951, Top Secret, PHST, SMOF–Naval Aide, box 24, State Department Briefs File, HSTPL.

61. Meaningful China Trade Control Differential, p. 4.

62. Ibid., p. 5.

63. Ibid., pp. 5–7.

64. Ibid., p. 2.

65. John Foster Dulles, memorandum to President Eisenhower (Secret), June 19, 1953, DH 1, Dulles John, June 53 (1), DDEPL.

66. JFDP, April 23, 1953, reel 204/205, 88971, Princeton Univ.

67. Chiang Kai-shek to Dwight D. Eisenhower, April 15, 1953, DH 1, Dulles April 53, DDEPL.

68. John Foster Dulles, telephone call with Dwight D. Eisenhower, July 23, 1954, DDE Diary Series 7, Phone Calls, June–Dec. 1954 (3), DDEPL.

69. John Foster Dulles, "Memorandum of Telephone Conversations" with Arthur W. Radford, A. C. Wilson, and Eisenhower, July 25, 1954, Dulles, J. F., Tel. Conv., box 10, July–Oct. 1954 (2), DDEPL.

70. President's comments to Dulles, July 26, 1954, DDE Diary Series 7, Phone Calls, June–Dec. 1954 (3), DDEPL.

71. Memorandum of Conversation with Sam Watson, Member of British Labor Party Delegation to Moscow and Peking, September 6, 1954, DH 4, Dulles Sept. 54 (2), DDEPL.

72. "China: Interference with British Merchant Shipping (Secret)," 1955, ADM 116/6245, TNA.

73. Eisenhower to Churchill, March 22, 1955, p. 3, Top Secret, DDE Diary Series 10, DDE Diary March 1955 (1), DDEPL.

74. Harry S. Truman to Pearl S. Buck, June 13, 1946, PHST, Official File, OF 148-A, box 757 (summarized copy in OF 220, box 964), File O.F. 150 (1945–46), HSTPL.

75. Statement by the President, November 30, 1950, PHST, Official File, OF 150, box 758, File O.F. 150 (1950–53) [1 of 2], HSTPL.

76. Conversations with Malcolm Muir about Colonialism, May 25, 1955, DDE Papers 5, ACW Diary May 1955 (2), DDEPL.

77. Preliminary Draft of Possible Statement of Position for Communication to the Republic of China, April 7, 1955.

78. Edward J. Marolda, "The U.S. Navy and the Chinese Civil War, 1945–1952" (PhD diss., George Washington Univ., Washington, DC, 1990), p. 372.

79. Notes Dictated by the President Regarding His Conversation with Sir Anthony Eden, Held Sunday, July 17, in the Afternoon, July 19, 1955, DDE Diary Series 11, DDE Diary July 1955 (1), DDEPL.

80. "Strategic Implications of the Application of Economic Sanctions against China," annex (Top Secret), n.d. [most likely February 1951], FO 371/92276, TNA.

81. A. A. E. Franklin, "Control of Exports from Hong Kong to China," January 15, 1951, FO 371/92274, TNA.

CHAPTER THREE

The First Taiwan Strait Crisis, 1954–1955

Before sending troops [to battle], first plan a path of retreat.
(未曾出兵先籌敗路.)

By 1953 the Korean conflict had deadlocked, and negotiations to end it had stale-mated. Stalin's death on March 5, 1953, helped break the negotiating logjam. A Korean armistice was finally signed on July 27, 1953. Military pressure from the south, exerted by Taiwan through the offshore islands, had been crucial in bringing it about. Soon afterward, however, the PRC began to redeploy troops from north to south, where tension gradually grew, particularly in the Taiwan Strait region. Two major crises would result, after the first of which Secretary of State Dulles would describe the situation as equivalent to "living over a volcano."[1]

On August 11, 1954, Zhou Enlai declared that the PRC must liberate Taiwan. On September 3, PLA forces began to bombard the offshore islands of Jinmen and Mazu. This renewed attempt to reclaim the islands by force showed Mao Zedong's independence from Moscow, but another important PRC goal was to interrupt the Nationalist block-ade. In November PRC leaders argued to Jawaharlal Nehru, the visiting Indian prime minister, that the Nationalists had been conducting "nuisance raids and interference with shipping" from their offshore bases. Upon his return, Nehru warned the British high commissioner in India that China was "determined not to tolerate this situation any longer."[2]

The shelling of Jinmen and Mazu was used as a cover to attack other Nationalist-controlled islands as well, in particular the most northerly of them, the Dachens. In this overcharged political climate, the U.S. government would sign a security pact with Taiwan and later, in January 1955, proclaim the Formosa Resolution. Ultimately the Americans would convince Chiang Kai-shek to evacuate the Dachens, and the U.S. Navy would provide the ships, training, and protection to do it. Thereafter the Nationalist-controlled offshore islands would encompass a much smaller area, from Jinmen in the south to Mazu in the north.

The Beginning of the 1954–1955 Taiwan Strait Crisis

Tension over the offshore islands led to a crisis during the fall of 1954. In August, Chiang Kai-shek ordered reinforcements to Jinmen and Mazu. The action had the outward appearance of preparation for a Nationalist invasion of mainland China. The PRC leaders were sufficiently concerned to authorize PLA attacks against the Nationalist-held offshore islands, and in particular the northernmost, Yijiang and the Dachens. Since late 1953 the Nationalists' position on the offshore islands had been more vulnerable than the year before; they now held only twenty-five, down from thirty-two.[3]

On May 22, 1954, a lengthy meeting was held between Eisenhower and his top advisers, including John Foster Dulles and his brother Allen W. Dulles (Director of Central Intelligence), plus Rear Adm. George W. Anderson Jr. (then Special Assistant to the Chairman of the Joint Chiefs of Staff), Adm. Arthur W. Radford (the chairman), and Robert Cutler (the national security adviser). The main topic was what attitude to take toward the offshore islands. In June 1950, the U.S. government had promised to defend Formosa, plus the Penghu Islands. However, no such guarantees had been extended to Jinmen Island. Asked by the *New York Times* on July 25 whether Jinmen was included, the State Department press officer, Michael J. McDermott, said no. A day later, the Nationalist ambassador, Wellington Koo, met with Dean Rusk (Second Assistant Secretary of State for Far Eastern Affairs) and Dulles and confirmed that the United States "would *not* help to fight off Communist attacks on the 'Quemoy islands.'"[4]

However, Admiral Radford "pointed out that it might not be necessary to report . . . an attack" by the Communists. U.S. airplanes, he acknowledged, were "quite frequently attacked . . . yet neither side had reported these events." After the meeting was over, it was clear exactly what the policy of air control would entail:

> If our Fleet on a patrolling mission, such as has been described, or if engaged in defending outlying islands against attack, was drawn into conflict with the Chinese enemy attacking such islands, our planes would not be justified in striking at targets on the Chinese Mainland. If, however, the Chinese Reds made an attack from the mainland on our carrier fleet, perhaps 100 or more miles at sea, then our security would permit us to follow such an attack in hot pursuit to the mainland bases.

Per the president's orders, however, U.S. forces would not "offensively attack the mainland of China in defending Formosa and the outlying islands, unless the security of our forces should require such an attack."[5]

Between June and September 1954, a series of top-secret reports by James A. Van Fleet (a recently retired general who had commanded the Eighth Army and UN forces in Korea, 1951–53) went to President Eisenhower. Van Fleet's assignment was to visit the East Asian countries allied with the United States to assess their political, economic, and military readiness to fight communism. On June 3 he sent his "preliminary" findings. This thirteen-page document examined the current American policies in the Far East;

offered assessments of Japan, Korea, Okinawa, and Formosa; made some brief observations; and laid out a plan for future work. According to Van Fleet, at the heart of the American national security policy in the Far East was the protection of "South Korea and the offshore island chain (Japan, the Ryukyus, Formosa, Philippines, Australia and New Zealand) from attack by communist forces." This did not include "offensive action forcibly to overthrow the Chinese Communist regime or to unite Korea," although, he noted, Washington did indirectly support "raids against the mainland and seaborne commerce of Communist China."[6]

Nationalist China's policy goals, Van Fleet reported, involved the "reconquest of all territories it formerly held or claimed (including ultimately Tibet, Sinkiang and Outer Mongolia) through the maximum mobilization of Chinese manpower on the island, with U.S. air, naval and logistic support." The Nationalists expected that although their own strength was inadequate, they could obtain a foothold on the mainland, at which point "wholesale defections of Communist troops" would allow them to return to power. Strategically, one of Taiwan's strongest assets, and the one Van Fleet mentioned first, was the array of offshore islands, collectively "a vital link in the chain of offshore island bases. Offensively, [it was] the single most important springboard for the launching of all manner of operations against the mainland of China." Were the Nationalist forces to be trained for such an operation, Van Fleet estimated, "about two years" would be required for the full development of their "military potential."[7]

Beijing was fearful of that prospect. On September 3, 1954, having relocated hundreds of thousands of troops from north to south following the reduction of tension in Korea, the PRC began to shell Jinmen Island. Two American military personnel were killed during the barrage, and another fourteen had to be evacuated. Acting Secretary of Defense Robert B. Anderson reported by top-secret message to President Eisenhower, who was at the summer White House in Denver, Colorado, that the advice of the JCS was split. The chairman and the service chiefs of the Air Force and Navy recommended that "National Policy should be changed to permit U.S. Naval and Air Forces to assist in the defense of 10 selected off-shore islands. Quemoy is included among the 10." The rest of the JCS, however, felt that "the Nationalist held off-shore islands near the mainland are important but not essential to the defense of Formosa from a military standpoint." Anderson warned that if the situation escalated and the Navy and Air Force were asked to assist the Nationalists to retain the offshore islands, doing so would, "in all probability, require some action by U.S. Forces against selected military targets on the Chinese mainland." As a short-term measure, the "CINCPAC [or CinCPac, the Commander in Chief, U.S. Pacific Command] has been alerted and directed to move carrier forces into a position from which support could be rendered, or a rescue mission undertaken, if directed."[8]

The Communist attacks on Jinmen also led to a series of discussions on whether the offshore islands could be held: Radford said yes; Gen. Matthew B. Ridgway, Chief of Staff of the Army, said no. What made intervention difficult was that Jinmen was in easy artillery range from the shore, so close that it would be "impossible for our vessels to maneuver between island and mainland." Intervention of any kind implied therefore a substantial U.S. commitment, which meant, as Eisenhower acknowledged, that American "prestige [would be] at stake. We should not go in unless we can defend it [Jinmen]." The president's "hunch [was] that once we get tied up in any one of these things our prestige [read: "face"] is so completely involved."[9] Anderson assured the president that his orders to the U.S. forces approaching the area were clear: as soon as they arrived they were to "get into position for reconnaissance and not to be aggressive." Eisenhower, in a relieved tone, replied, "We are not at war now."[10]

On September 8, the CIA produced a thirty-three-page, top-secret assessment of the situation, "The Chinese Offshore Islands." It described how in just five hours on September 3, approximately six thousand 120 and 155 mm artillery shells had been fired at Jinmen, Little Jinmen, and Nationalist ships at anchor. Accuracy was high, owing, according to unsubstantiated reports, to spotting by "Communist agents on Quemoy." The U.S. and ROC reaction was rapid, and by the morning of September 5 there were "three carriers, a cruiser and three destroyer divisions of the Seventh Fleet . . . patrolling the waters in the Formosa Straits at a distance of several miles from Quemoy." Nationalist naval and air counterattacks against mainland China during September 6–7, 1954, reportedly destroyed five out of the fourteen guns being used by the Communists.[11]

The CIA study listed Nationalist troop strength (thirty-four thousand on Jinmen and another 4,500 on the neighboring island of Little Jinmen) and detailed its disposition: troops, artillery, and mortars, plus air, naval, and logistical support. The Nationalists' naval assets were particularly important, to protect future resupply convoys. Of the Nationalist navy's total of "56 major vessels," the CIA estimated what it could task to defend Jinmen as "2 destroyers, 3 destroyer-escorts, 2 patrol gunboats, 2 smaller patrol vessels, and about 10 armored junks." Taiwan had about 850 aircraft, of which 415 were combat aircraft; most were "obsolete World War II" planes that could only be used in daylight: the Nationalist air force's "capacity for night interception and strategical bombing is nil." The CIA estimated that fifteen days' supplies were already on the islands and that Taiwan could provide another forty-five days' worth if required.[12] All in all, the CIA report had described a dire situation of crisis proportions. A series of NSC discussions on how best to respond to it soon followed.

The National Security Council Discussions

On September 12, 1954, the NSC submitted a top-secret paper, divided into three sections: prowar considerations, antiwar factors, and recommendations. On the prowar side it argued that "Quemoy cannot be held *indefinitely* without a general war with Red China in which the Communists are defeated." The PRC might grant Taiwan its independence but would never allow the "alienation of the off-shore islands like Quemoy." Therefore, if Eisenhower wanted a war with China, "Quemoy can be made to provide the issue." However, Congress would almost certainly oppose any such action, in particular since "for four years" the various offshore islands "have not been included in the area the [Seventh] Fleet is ordered to defend." Also, war with China would "alienate world opinion and gravely strain our alliances, both in Europe and with Anzus."[13]

As for the no-war option, the NSC predicted—quite accurately, as it would turn out— that "it does not seem that any all-out Chicom [Chinese Communist] assault [of the offshore islands] is likely in the near future because of (a) early adverse weather conditions; and (b) uncertainty as to US reaction." The longer the United States aided the Nationalists, however, the greater would be "the loss of US prestige"—again, just another word for "face"—"if the Island is later lost while the US stands by." Therefore, the basic problem, "if we want to avoid all-out war with China, is to do so [i.e., aid the Nationalists] on terms that will avoid a serious loss of Chinat [Chinese Nationalist] morale and US prestige."[14] Later, the Department of State's Policy Planning Staff specifically warned Dulles that Washington was "drifting by degrees into a situation in which our prestige will be committed to the retention of the islands."[15]

On April 5, 1955, Eisenhower warned Dulles against committing "United States military prestige to a campaign under conditions favorable to the attacker."[16] Dulles agreed with Eisenhower on the problems of "losing face" in the defense of the islands:

> It is in the interest of the Republic of China, as well as in our own, that the Republic of China not commit its prestige to the defense of these perhaps indefensible positions so deeply that, if they should be lost, all future possibilities now represented by the Republic of China would also be lost. The lesson of Dien Bien Phu [in Vietnam, lost by the French to the Viet Minh after a three-month defense in 1954] should not be forgotten. Originally conceived to be an outpost of transitory value, it gradually became converted into a symbol, so that when it fell, all else fell with it. The same mistake should not be repeated in regard to Quemoy and Matsu, islands which without American aid are probably indefensible, and even with it may not be defensible except by means which would defeat the large common purpose.

Eisenhower concluded from all this that "militarily and politically we and the Chinese Nationalists would be much better off if our national prestige were not even remotely committed to the defense of the coastal islands."[17] One way to shield American prestige was to turn to the United Nations.

Requesting United Nations Intervention

An upshot of the NSC discussions was a decision to "explore the possibility of a US appeal to the [UN] Security Council." The purpose would be to "charge the Chinese Communists with aggression against Quemoy and threatened aggression against Formosa,"

which would be a breach of article 39 of the United Nations Charter. As the NSC
pointed out, "action against Quemoy is avowedly part of a program to take Formosa by
force," making the situation "*not* purely domestic, civil war."[18] With UN support, a demand could be sent to the Chinese communists to "refrain, for a stipulated period, from
military action directed against areas which have been held by the Republic of China
since the close of the Second World War." But if no action were taken, the Nationalists
might have to withdraw from the offshore islands, which would be a "serious blow to the
morale of the Chinese Nationalists on Formosa and to the prestige of the United States
in the Far East, and it would raise the question as to whether the US was really prepared
to stand anywhere in the East."[19]

Attempts by Washington to get the United Nations involved had in fact begun immediately after the shelling of Jinmen. For example, Dulles met in London with Anthony
Eden, then foreign minister, and the New Zealand high commissioner to see whether
the Security Council could meet and, under Chapter VI of the UN Charter, order the
PRC to suspend its attacks on the offshore islands. New Zealand was a member of the
Security Council, and its being an Asian country gave "it a legitimate interest in the
situation." These plans were being carried out with the "utmost secrecy."[20]

One major consideration in working through the UN was hope of undermining Russian support for China. Asking for a vote under article 40 might split apart the USSR
and PRC, since if the Soviets vetoed the move, they would impair their ongoing "peace
offensive." But if the "Soviets did *not* veto, the Chicoms could react adversely," perhaps
even defying the UN and becoming an "international outcast."[21]

In addition to giving the United States leverage over China, turning to the UN might
have other subsidiary effects, some negative but others potentially quite positive for
Washington. First, if the Communists were restrained in their attacks on the offshore
islands, the Nationalist attacks on the mainland might also end. Second, less favor-
ably, UN action "might end the embargo on Red China to the extent that it exceeds the
restrictions against strategic goods to the Soviet Union." Third, however, if the Communists rejected a UN demand, the moral "sanction" of the "free world against the Communist world would be reinstated," whereas if they accepted it one "probable ultimate
outcome of UN intervention, if the Soviet Union permitted it, would be independence of
Formosa and the Pescadores [Penghus]."[22]

On September 29, 1954, Walter Bedell Smith (a retired general, "Ike's" chief of staff
through much of World War II, now under secretary of state) called the president to tell
him that the State Department backed the proposal that New Zealand should present
the case under Chapter VI, and the "President agreed."[23] On October 5, 1954, Eisenhower told Dulles to be careful not to discuss it with "anyone who might allow it to leak

to the public," that only just before the UN meeting should they "notify Senator [William F.] Knowland [R-CA, the Senate majority leader] of our position."[24] On October 12, 1954, Dulles could report to the president: "After difficult negotiations, we have a very complete and definite understanding with the United Kingdom and New Zealand."[25]

Taiwanese support would be crucial if the UN intervention was to succeed. On December 18, Eisenhower met with the Taiwanese foreign minister, Dr. George C. Yeh. As anticipated, Yeh asked for a continuation of American logistical support for "the defense of the off-shore islands." In response, Eisenhower referred to a recent NSC decision to "for the present, seek to preserve, through United Nations action, the status quo of the Nationalist-held off-shore islands; and, without committing forces except as militarily desirable in the event of Chinese Communist attack on Formosa and the Pescadores, provide to the Chinese Nationalist forces military equipment and training to assist them to defend such off-shore islands, using Formosa as a base."[26]

New Zealand duly made the desired proposal to the UN, and the PRC flatly rejected it. Zhou Enlai even objected that "if the New Zealand idea was to discuss these islands, then that constituted intervention in China's internal affairs." The British chargé d'affaires in Beijing, Humphrey Trevelyan, tried in a two-and-a-half-hour interview to persuade Zhou to let the Security Council discuss the dispute, assuring him that the "New Zealand proposal would be merely that the Security Council should discuss the serious situation existing in the area of the islands, since this was the immediate point of danger." Eisenhower promised that the "Chinese Government should be present at the discussion." Zhou turned all of these offers down, reaffirming that the disputed islands "were Chinese, and the Chinese would liberate them."[27]

In late February 1955, Eisenhower warned Dulles that if the UN did not sponsor a cease-fire soon, it might be difficult to stop Chiang Kai-shek from taking action. At Dulles's suggestion, Eisenhower decided it was best to tell the British government that time was running short:

> I agree that it would be wise to inform Eden that unless we soon arrive at a cease fire, we cannot much longer insist that the present policy be observed which permits major Communist build up or attacks without Chiang reaction. I believe that you should tell him that we do not intend to blackmail Chiang to compel his evacuation of Quemoy and the Matsus as long as he deems their possession vital to the spirit and morale of the Formosan garrison and population. On the contrary we expect to continue our logistic support of Chiang's forces as long as there is no mutually agreed upon or tacit cease fire. Finally, if we are convinced that any attack against those islands is in fact an attack against Formosa, we should not hesitate to help defeat it. Possibly you should tell him too that because of the continuing build-up of Chicom forces, we cannot tell when any of these emergencies might arise.

But Eisenhower made clear that the authority to order American troops to assist Chiang was his and his alone: "Any offensive military participation on our part will be only by order of the President."[28]

On February 25, 1955, Dulles reported that Eden had convinced him it was best not to continue to press for a cease-fire in the UN. Rather, Eden would try to convince the PRC to agree not to "seek a violent solution of Formosa matter." Eden's draft message to Beijing read:

> I, therefore, wish to inquire whether the Chinese Government would state, privately or publicly, that, while maintaining their claims, they do not intend to prosecute them by force. If they were prepared [to] give such assurance, we would be ready so to inform the US Government and approach them with, *we believe, a good hope* of finding basis for a peaceful settlement of the situation in the coastal islands.

To this Dulles had replied "that we would agree to suspend for a further brief period [our] request for a cease-fire resolution so as to permit this other initiative of his to have [the] best chance of success."[29] Washington did not expect Taipei to support a cease-fire resolution. Accordingly, in discussions in Taiwan in early March with Chiang Kai-shek, Dulles urged, "I must, however, request with the greatest possible strength and earnestness that if such a resolution came to a vote he [Chiang] would not veto it but would let the Soviets veto it." Chiang laughed and agreed that "actually he would prefer to let the Soviets veto it."[30]

Soon thereafter, during a trip to Canada, Dulles told the cabinet in Ottawa that Washington had convinced Chiang Kai-shek to accept the following four points:

1. Limited our Treaty to Formosa and the Pescadores.

2. Secured his agreement not to attack the Mainland from any position without our prior agreement.

3. Secured his evacuation of the Tachens and Pishan.

4. Sought a cease-fire for the Formosa Straits which he felt would be a death blow to his hopes.

Asked by the Canadians what further concessions Chiang could be asked to make, Dulles warned, as he reported later, "I did not know whether we could make him take any more bitter medicine at this time without disastrous consequences."[31]

It was not only in the United Nations that Washington was attempting to negotiate a cease-fire in the Taiwan Strait. In April 1955, Dulles also worked with a number of African and Asian countries, most of them newly independent, attending a major conference in Bandung, Indonesia. These delegations duly pressured Zhou Enlai, pushing him, according to Dulles, to "follow a pacific rather than belligerent course." In late April Dulles told a number of American senators, including Knowland, Bourke B. Hickenlooper (R-IA, strongly conservative), and Margaret Chase Smith (R-ME), that "we should be prepared to talk with the Chinese Communists merely to the extent of ascertaining whether they would make a 'cease-fire.'" Were the opportunity to arise at Bandung, "a complete turn-down by the United States would alienate our Asian non-Communist friends and allies." To this, Senator Knowland responded that "we could not trust a cease-fire agreement and that the Armistices in Korea and Indochina were already being broken."[32] At about the same time, Dulles suggested to the Soviet foreign minister, Vyacheslav Molotov, that a six-power conference be convened to discuss the Taiwan Strait, so as to create "a situation where as in Germany, Korea and Vietnam, it was agreed that unification would not be sought by force."[33]

During the late spring, the PRC sent to the United States, by means of Pakistan, India, and others, offers to negotiate. For example, in early May a Pakistani, Mir Khan, told Henry Cabot Lodge Jr., then ambassador to the United Nations, that in his opinion "the Chinese Communists want to climb down, that they are willing to acknowledge the present status quo in the Formosa area for ten or twenty years." On this view, Beijing was now willing to release eleven American airmen it was holding—and, importantly, not as a tit for tat for any side agreement to allow the PRC into the United Nations.[34] In June 1955, U Nu, prime minister of Burma, relayed a message that the Chinese communists were ready to "sit down directly and negotiate a cease-fire."[35]

In addition, Zhou Enlai told India's ambassador to the UN, Krishna Menon, that the PRC could not agree to a formal cease-fire with Taiwan until there "was evidence of substantial withdrawals" of Nationalist troops from Jinmen and Mazu but that once there was, "there need be no particular time limit for the standstill in hostilities." The Nationalists would not need to "surrender" the islands to the Communists, only not pursue there a "scorched earth policy such as had been carried out in the Tachens."[36] By mid-July, Menon had visited Eisenhower two times to ask whether talks could be established with the PRC. In return for talks, Menon was sure, captured American airmen in China would be released; nevertheless the president held firmly that "the American people will not consider using the lives and freedom of their own citizens as a bargaining" chip, that Washington would not be willing to deal with Beijing "in good faith until *after* they have released these prisoners."[37] On July 18, 1955, John Foster Dulles sent to Eisenhower a list of the thirty-eight American civilians and eleven airmen being held by China.[38]

Hearing signals from at least three sources, leaders in Washington "agreed this is a typical Chinese trick—they [i.e., the Chinese] will take whoever has the most to offer—which[ever] negotiator is most successful."[39] However, as lengthy (extending throughout March and April) and largely unproductive negotiations with him had shown, Chiang Kai-shek was determined not to withdraw from the two offshore islands in question under any circumstances. The U.S. government now saw itself obliged to increase its military aid to Chiang, and largely on his terms.

Discussions on New U.S.-ROC Security Arrangements

The immediate political impact in Washington of the Communist attack on Jinmen, then, was not what Beijing expected. Rather than push the United States and Taiwan farther apart, this crisis resulted in renewed movement toward a new security treaty.

As early as June 7, 1953, Chiang Kai-shek had mentioned to President Eisenhower the possibility of "bilateral or multilateral security pacts" between the United States and East Asian countries aside from Taiwan, notably South Korea, Thailand, and those of Indochina.[40] As for a U.S.-ROC pact, however, all previous discussions had ended in failure. To this point, U.S. Navy operation orders had specified that the term "enemy forces" meant those attacking Taiwan or the Penghus, not other offshore islands held by the Nationalists.[41]

Washington had to tread softly. Public opinion in the United Kingdom opposed war with China. Churchill had privately warned Eisenhower that "a war to keep the coastal islands for China would not be defensible here."[42] The British ambassador, for his part, emphasized to his interlocutors that the United Kingdom had formally recognized the PRC and "that these [offshore] islands were part of China." This made it highly unlikely that Britain could support the United States in any fight over the offshore islands; should war break out, "the Western alliance might be split. . . . Were these islands," the ambassador asked, "really worth it?"[43]

Taking into consideration the differing views of its allies, Washington adopted an intentionally ambiguous policy toward the pact negotiations, one that, while reaffirming the pledges of Truman and Eisenhower to defend Taiwan and the Penghus, would keep the Communists unsure about the true defense status of the offshore islands.[44] A pact on these lines would give the president maximum flexibility, laying down "no unconditional Presidential decision to defend the coastal positions."[45] According to one press report, "The pact will be deliberately vague about how the United States might react if the Reds were to invade any of the other Nationalist-held islands off the China coast. The U.S. doesn't want the Reds to know which it will defend, and which it will simply write off. It prefers to keep them guessing."[46]

After lengthy negotiations, the two sides agreed to very specific draft language: that the American security guarantee to the Republic of China was also "applicable to such other territories as may be determined by mutual agreement."[47] This wording deferred any decision on whether the offshore islands would be included in the treaty. Meanwhile, a top-secret State Department report argued that a mutual-security treaty should cover "Formosa and the Pescadores, but not the offshore islands." The treaty should be "defensive in nature," similar to the German renunciation of "force to unite Germany" or to the pact with South Korea, where the U.S. government opposed "the use by [President Syngman] Rhee of force to unite Korea."[48]

The U.S.-ROC Mutual Defense Treaty was signed on December 2, 1954, with this ambiguous wording. Interestingly, prior to China's attacks the U.S. government had hesitated to finalize this treaty, because of the "inherent difficulty of concluding a purely defensive pact with a country engaged in actual hostilities," especially since Washington hoped to "avoid direct United States involvement."[49]

Eisenhower, however, insisted on an exchange of separate notes on veto power over any Nationalist attack. Washington did not want Chiang calling the shots, having the power to declare war. A secret agreement was signed to make sure this would not happen. As Dulles explained on December 22, 1954, to the White House press secretary, "there is an exchange of notes which more or less gives us the power to control any offensive operations from Formosa, etc."[50] Two months later, Eisenhower told Churchill that in this secret agreement Chiang had promised he would not "conduct any offensive operations against the mainland either from Formosa *or from his coastal positions,* except in agreement with us." In addition, by the agreement the Nationalists were limited in "their right to take away from Formosa military elements, material or human, to which we had contributed if this would weaken the defense of Formosa itself."[51]

Chiang Kai-shek confirmed this promise in a March 4, 1955, discussion with Dulles: "Now that the Mutual Security Treaty was in force," Dulles later recorded, "he wished to assure me that he would take no independent action insofar as the use of force was concerned, and would undertake no large-scale military operations against the Mainland without full consultation with the US."[52] Further, as Eisenhower explained to Churchill, the secret agreement would allow Washington to stop Chiang from using his offshore bases to continue the "sporadic war against the mainland" or to support an "invasion of the mainland of China." Eisenhower felt that the secret agreement with Chiang showed that Washington had "done much more than seems generally realized."[53] One thing it had accomplished, and one of the most important concessions Eisenhower convinced Chiang to make, was the evacuation of the Dachen Islands.

The Evacuation of the Dachens

The Dachens (or Tachens) were Nationalist-controlled islands

20

The Secretary of State to the Chinese Minister of Foreign Affairs

DEPARTMENT OF STATE
WASHINGTON
Dec 10 1954

EXCELLENCY:
 I have the honor to refer to recent conversations between representatives of our two Governments and to confirm the understandings reached as a result of those conversations, as follows:

 The Republic of China effectively controls both the territory described in Article VI of the Treaty of Mutual Defense between the Republic of China and the United States of America signed on December 2, 1954, at Washington and other territory. It possesses with respect to all territory now and hereafter under its control the inherent right of self-defense. In view of the obligations of the two Parties under the said Treaty and of the fact that the use of force from either of these areas by either of the Parties affects the other, it is agreed that such use of force will be a matter of joint agreement, subject to action of an emergency character which is clearly an exercise of the inherent right of self-defense. Military elements which are a product of joint effort and contribution by the two Parties will not be removed from the territories described in Article VI to a degree which would substantially diminish the defensibility of such territories without mutual agreement.

 Accept, Excellency, the assurances of my highest consideration.

JOHN FOSTER DULLES
*Secretary of State of the
United States of America*

His Excellency
 GEORGE K. C. YEH,
  *Minister of Foreign Affairs of
  The Republic of China.*

TIAS 3178

24

English Text of Foregoing Note

DECEMBER 10, 1954

EXCELLENCY:

I have the honor to acknowledge the receipt of Your Excellency's Note of today's date, which reads as follows:

"I have the honor to refer to recent conversations between representatives of our two Governments and to confirm the understandings reached as a result of those conversations, as follows:

"The Republic of China effectively controls both the territory described in Article VI of the Treaty of Mutual Defense between the Republic of China and the United States of America signed on December 2, 1954, at Washington and other territory. It possesses with respect to all territory now and hereafter under its control the inherent right of self-defense. In view of the obligations of the two Parties under the said Treaty and of the fact that the use of force from either of these areas by either of the Parties affects the other, it is agreed that such use of force will be a matter of joint agreement, subject to action of an emergency character which is clearly an exercise of the inherent right of self-defense. Military elements which are a product of joint effort and contribution by the two Parties will not be removed from the territories described in Article VI to a degree which would substantially diminish the defensibility of such territories without mutual agreement."

I have the honor to confirm, on behalf of my Government, the understanding set forth in Your Excellency's Note under reply.

I avail myself of this opportunity to convey to Your Excellency the assurances of my highest consideration.

GEORGE K C YEH
*Minister for Foreign Affairs of
the Republic of China*

His Excellency
JOHN FOSTER DULLES
*Secretary of State of
The United States of America*

to the far north, in Zhejiang Province. Beginning in mid-May 1954, the Communists occupied six small islands near the group. A PRC invasion appeared to be planned, so the Seventh Fleet was sent there in early June and again in August in shows of force. Nevertheless, as the CIA warned in September 1954, there were at least three Chinese armies within 150 miles of the Dachen Islands, any one of which was capable of taking the islands if it could coordinate naval and air support.

The Dachen Islands were over two hundred miles from Taiwan, a distance that gave the PRC a clear strategic advantage: "The Tachens are particularly vulnerable to air attacks by large numbers of aircraft of all types" that were based at a number of mainland airfields in Ningbo, Hangzhou, and Shanghai.[54] The distance also meant, however, that the PLA threat to them was more symbolic than strategic: to stage an invasion of Taiwan from the northern offshore islands did not make any military sense—they were simply too far away. Rear Adm. Samuel Frankel of the Office of Naval Intelligence argued that "from the viewpoint of protecting Taiwan, I think that these islands have no significance at all."[55] As early as July 15, 1953, Adm. Arthur W. Radford, at the time "dual hatted" as the Commander in Chief, U.S. Pacific Command and Commander in Chief, Pacific Fleet, having warned Chiang Kai-shek to "bolster the sagging defense" of the northern islands, reported back to Washington that Chiang was "reluctant to deploy adequate forces." Radford's concern was that the Dachens were psychologically important to Taiwan's defense and also strategically important for gathering intelligence on the PRC. For a time, Radford even considered proposing that these islands be put within the "U.S. defense perimeter."[56] On July 15, 1953, the Eisenhower administration concluded that "although the importance of these islands to Taiwan's defense is generally recognized here, the prevailing view is that the responsibility for their defense must remain with the Government of China."[57]

During November 1954, Prime Minister Nehru was told by Chinese leaders that the Nationalist blockade was a major problem, that they were "faced with continuous pin-pricks and irritations of cumulative effect and he had the definite impression that they were determined not to tolerate this situation longer."[58] In fact, on November 1 PRC planes had already begun to bomb and strafe the Dachen Islands. In early 1955 the PLA focused its attention on Yijiangshan (North and South Yijiang Islands), eight miles north of the Dachens. On January 18, 1955, over fifty PLA Air Force (PLAAF) planes attacked the islands, which finally fell on the 20th. On January 18, Dulles told the president that Yijiangshan was "at such a distance and is of such little importance, that we could not view this with any great concern." Dulles argued that losing these islands would "not affect vital interests of Formosa or ourselves." Eisenhower agreed, and they decided to call it "a skirmish of no significance whatsoever."[59] However, on January 19 the PLAAF began to attack the Dachen Islands as well, with seventy aircraft.[60]

These islands were closer to Taiwan and considered to have at least symbolic importance. This new attack required an American decision either to order the U.S. Navy to defend the Dachens or to convince the Nationalists to abandon them. The Nationalists requested that the Seventh Fleet be moved closer to the Dachens to expedite logistics. Later that month, they also asked for Navy air support. The American ambassador to Taiwan backed both requests, to avoid "undermining confidence in our strength and determination."[61] But defending the Dachen Islands permanently would be difficult, requiring two full-time aircraft carriers plus their supporting ships. Instead, Dulles affirmed the earlier view that the Dachens "were too far from Formosa, too vulnerable, and insufficiently important from the strategic point of view to justify an American commitment to defend them."[62]

On January 19, 1955, Secretary Dulles told the president and Admiral Radford over lunch that "doubt [in Asia] as to our intentions was having a bad effect on our prestige." He recommended that the Nationalists be encouraged to evacuate the Dachen Islands, in return for a public agreement to defend Jinmen Island. The American stance toward the Mazus, collectively, would be left unclear. Radford and the president agreed, after which Dulles drafted an aide-mémoire:

1. [It had been decided to] encourage the CHINATS to evacuate Tachen and the other offshore islands exclusive of Quemoy.

2. The United States would provide sea and air protection so as to permit of an orderly evacuation.

3. Contemporaneously, the United States would state that in view of the aggressive actions of the Chinese Communists and their proclaimed intention to seize Formosa, the United States will assist the CHINATS to hold Quemoy Island which, under existing circumstances, is deemed important for the defense of Formosa and the Pescadores.

In order to enact this decision, it was agreed, Dulles would contact the Chinese minister, then the British ambassador, and third the congressional leadership. Of that last group, Dulles wanted in particular to "ascertain whether Congress will extend the necessary authority to carry out the above course of action, which should be broad enough to permit of attacking the mainland about Quemoy, if that was deemed essential to prevent a buildup which would dangerously threaten Quemoy."[63]

On the 21st the president met again with Dulles and Radford, and this time the rest of the Joint Chiefs as well. Adm. Robert B. Carney, the Chief of Naval Operations (CNO), acting as the chiefs' spokesman, immediately pointed out the difficulty of evacuating thirty thousand people from the Dachens. Supported by Radford, Carney warned "the evacuation would be much more arduous than their [i.e., the Dachen Islands'] defense or reinforcement." However, the Joint Chiefs all agreed that the Chinese Nationalists "cannot defend Tachen," so in Dulles's view the choices were really between evacuation and surrender: "The issue is clear, you must try to get them out or let them surrender."[64]

On January 23, the U.S. government for the first time recommended to Chiang Kai-shek that the Dachens be evacuated. Chiang agreed to withdraw, but reluctantly, in particular because giving up such a strong position without a fight would "gravely affect troop and civilian morale." To avoid the impression that he was backing down generally, Chiang refused to enter into a cease-fire with the PRC of any kind.

On January 28, Admiral Radford visited President Eisenhower at the White House to talk about the Dachen situation. The most important issue was the operational order the Chief of Naval Operations would need to send to the commander in chief in the Pacific, Adm. Felix Stump. The CNO operation order was to have four points:[65]

a. It was agreed that if called upon by the Chinese Nationals, the American forces would assist in the evacuation of the Tachens.

b. It was further agreed that if any attack was made against this operation, that the American forces were, of course, fully authorized to defend themselves as necessary.

c. It was agreed that there would be no attack on Chinese bases unless this was essential to the success of the operation. It was further agreed that if such attacks became necessary, they would be carried out only against air fields positively identified as contributing forces to the attack against us.

d. Finally, it was agreed that the Commander in the area would authorize no attacks against the Chinese mainland on any initial sortie by the ChiComs. It would first be determined by the Tactical Commander that the purpose of the ChiComs was to continue the attacks before this type of action would be undertaken. However, it was to be clearly understood that if the ChiComs undertook a consistent and persistent air attack against the operation, that the United States forces would be authorized within the limits of the above stated to take such action as was essential to protect themselves and to assure the success of the operation.

The wording of this guidance to the CNO, especially the fourth point, suggested that strikes against a variety of targets on the Chinese mainland might be delivered in response to a Chinese attack. On January 29, Radford made clear to CinCPac, Adm. Felix Stump, that U.S. attacks on mainland China could be authorized even if not "essential to success . . . if [they proved] necessary in defense of own forces engaged in the operation."[66] Two days later, however, the permissible targets in such situations were limited to airfields: "It was finally agreed that the United States Commander could attack the airfields from which the Chinese Communists air forces were operating if necessary in defense of his own forces engaged in the operation."[67]

At the same time as this order to CinCPac, duly forwarded by him, Commander in Chief, Pacific Fleet issued an operations plan for the defense of the evacuation. Ships were not to sail within three miles of China's coastline and were if "possible [to] avoid shooting incidents." But the "commanders and pilots should be instructed not [to] accept positions of tactical disadvantage in interest [of] avoiding" shooting incidents. If the PLA attacked the Nationalists, U.S. forces could "provide naval air and gunfire support." Finally, if combat broke out, "hot pursuit into mainland air space or territorial waters is authorized for purpose of successful conclusion of engagement."[68]

The evacuation of the Dachens was a massive undertaking. Between February 7 and 11, 1955, the U.S. Navy used a total of 132 ships and four hundred aircraft to evacuate 14,500 civilians, ten thousand Nationalist troops, and approximately four thousand guerrilla fighters, along with over forty thousand tons of military equipment and supplies. To protect the evacuation, the Seventh Fleet assembled a "backbone" of six aircraft carriers. China was warned not to interfere, that U.S. Navy forces had "instructions not to provoke incidents but they also have instructions not to accept any tactical disadvantages." Or, put another way, "American airmen were not to get 'altruistically shot down.'"[69]

The international reaction was mixed. The Australians compared the loss of the Dachens (which the PLA occupied on the 13th) to the 1938–39 fall of Czechoslovakia and the PRC threat to invade Taiwan to the 1939 invasion of Poland. In a February 1955 Gallup poll, 66 percent of Australians "declared themselves in favour of Australia joining the United States in any war which might result from United States efforts to prevent Chinese Communists from invading Formosa." In April Prime Minister Menzies argued the "desirability of giving Formosa military and political strength to ensure the future will be decided peacefully and not as a result of Communist policies of force."[70]

The decision to evacuate the Dachens gravely undermined Nationalist morale. In February 1955, Admiral Stump and Dulles emphasized during American-British talks the importance of that factor:

> Admiral Stump explained to Eden defense relationship between offshore islands and Taiwan. They block launching attack on Taiwan, provide advance warning and are closer to hostile area in case of fighting. Field Marshal [John] Harding [Chief of the Imperial General Staff] interjected to differ with Stump. Comparing situation to Allied assault in Operation Overlord [the 1944 Normandy landings], Harding expressed opinion critical question is not launching or lodging initial attack across water but in being able afterward sustain assault forces. He thought Chinese Communists military leaders would advise against attack on Taiwan as long as Seventh Fleet commanded sea and air. Hence he did not (repeat not) believe possession offshore islands would have much to do with whether Chinese Communists would or would not attack Taiwan.[71]

During a July 17 meeting between Eisenhower and Eden, Ike emphasized the symbolic value to Chiang of the offshore islands. He warned that "another single backward step in the region would have the gravest effects on all of our Chinese friends," and after the meeting was able to report that Eden "had no trouble understanding the importance of morale in Chiang's army on Formosa."[72]

After the evacuation, the flag of the Republic of China was lowered in the Dachen Islands by Chiang Ching-kuo, Chiang Kai-shek's son. The ROC's provincial government apparatus for Zhejiang (in which the Dachens lay) was abolished; this meant that Nationalist forces now held disputed territory only in Fujian Province. Rather than pushing the United States and Taiwan farther apart, however, as Beijing had undoubtedly hoped would happen, greater cooperation leading up to the evacuation of the Dachens led to closer relations between Washington and Taipei. The possible use of force to protect the other offshore islands had by that time been detailed in the "Formosa Resolution."

Negotiating the Formosa Resolution

PRC attacks on the Dachen Islands during January 1955 had convinced Eisenhower to request from Congress special powers to defend Taiwan. It was decided to sign a second document, to extend certain American defensive guarantees concerning Taiwan to the offshore islands. This idea was perhaps first proposed by the Nationalist foreign minister in a December 20, 1954, meeting with President Eisenhower. During this meeting, Dr. Yeh quoted Chiang Kai-shek: "With respect to the off-shore islands, the Generalissimo recognized that the Treaty did not cover these, but felt that it would be a good psychological warfare move for the United States to give some form of assurance that it would provide logistic support for Chinese forces engaged in their defense." In reply, Eisenhower said that while Washington was not "indifferent as to these islands," it would be better to handle the offshore islands on a "case by case" basis, "each on its merits."[73]

On January 20, 1955, Secretary of State Dulles met with six senators, six congressmen, and the chairman of the Joint Chiefs of Staff, to discuss what he called "a sounder defensive concept" to handle tensions in the offshore islands.[74] Dulles explained that the president preferred evacuation of the Dachen Islands but thought the administration should have "some authority from Congress to use the armed forces of the United States in the area for the protection and security of Formosa and the Pescadores."[75] Dulles argued that if "the [i.e., a new] Formosan treaty were ratified, and the President were given these powers, there will be a realization that we have reached the point that we are not going to retreat more and it possibly will have a stabilizing effect." Not to take this action would, in his mind, make the "risk of war" greater.[76]

However, the important thing, as Eisenhower and Dulles agreed in a telephone conversation, was that the resolution should allow for possible "enlargement of the area described in the treaty."[77] On January 25, 1955, Eisenhower wrote a top-secret letter to Churchill responding to the British concern that the United States might get drawn into a "Chinese war":

> It is probably difficult for you, in your geographical position, to understand how concerned this country is with the solidarity of the Island Barrier in the Western Pacific. Moreover, we are convinced that the psychological effect in the Far East of deserting our friends on Formosa would risk a collapse of Asiatic resistance to the Communists. Such possibilities cannot be lightly dismissed; in our view they are almost as important, in the long term, to you as they are to us.

After reassuring Churchill that he was working for the "preservation and strengthening of the peace," Eisenhower added, "But I am positive that the free world is surely building trouble for itself unless it is united in basic purpose, is clear and emphatic in its declared determination to resist all forceful Communist advance, and keeps itself ready to act on a moment's notice, if necessary."[78]

In the January 28, 1955, meeting cited above between Humphrey Trevelyan and Zhou Enlai, the Chinese views became clear. A concise transcript organized them in six points. First, Zhou accused the president of sending a "war message" to Congress; second, he denounced Washington for using the UN as a cover "for aggression against China"; third, he asserted that the UN had no right to discuss the offshore islands, since they were a matter of "internal sovereignty"; fourth, he argued that the offshore islands dispute could not be separated from Taiwan's status; fifth, the PRC would never cut a "deal" on the offshore islands but planned to "liberate" them; and finally, the Chinese were "not afraid of war threats and would resist if war was thrust on them." Trevelyan described Zhou's attitude during their lengthy meeting as "tense and absolutely uncompromising."[79]

On the 29th the Formosa Resolution was passed by Congress.[80] It stated that only the president could judge whether a PRC attack on the offshore islands was part of a more general assault on Taiwan. This vague wording worried Chiang Kai-shek; that same day, Ambassador Karl L. Rankin telegrammed to Washington to report that a very "nervous" Chiang, presumably having seen the text approved by the House of Representatives on the 25th, had expressed hope for a specific statement of American intent to help defend Jinmen and Mazu. Rankin had replied that while he did not know the details, perhaps a vague resolution, without "mentioning them [the islands] by name," would have the "advantage of not implying all other islands being written off."[81]

Dulles had met with Yeh on the 28th. Yeh requested that a "joint announcement [on] Quemoy and Matsu" be issued publicly. The secretary informed the foreign minister that under the present circumstances, the United States indeed considered Jinmen and Mazu important to the defense of Formosa—but he did so only "orally and confidentially":

> Secretary said any formal Chinese statement should avoid implication of agreement or commitment between United States and Chinese Governments. United States responsibilities as to "related area" were unilateral. There was no agreement as to related area, and United States might have to deny any implications to contrary. Chinese unofficial sources could speculate, but there should be no official statement on either side as to understanding re defense of related area. Secretary felt impossible to draw absolute geographical line or to specify which islands were important to defense of Formosa and which were not important. Relative defensive importance of islands could change. Any significant Communist buildup in area would be regarded by United States with concern.[82]

It was best to be ambiguous, then, and not state this publicly, argued Dulles, since by the terms of the Formosa Resolution it was up to the president to decide what the term "related area" actually meant.

On the 30th, the day after the Formosa Resolution was signed, Ambassador Rankin warned of an extremely "difficult situation" with regard to Chiang Kai-shek. In what appeared to be an ultimatum, which was one of his standard negotiating tactics, Chiang had informed Rankin that "his government would not request assistance in withdrawing from Tachens [which was to happen the next month] until US position re Kinmen [Jinmen] and Matsu clarified. Failure to insist on this would betray China." Chiang's understanding of events told him that three things should now happen:

1. After approval of resolution by Congress, two governments would issue simultaneous and complementary statements on offshore islands.

2. Above statements would provide for US assistance in evacuation of Tachens.

3. At same time, it would be made clear US was extending protection to Kinmen and Matsu.

Chiang considered this a matter of "honor" and was willing to lose everything not to "lose face": "He said Tachen and forces there might be lost, that Formosa where he and his people were prepared to die might also be lost, but that if China's honor were preserved for posterity it would be worthwhile." Chiang reminded Washington that "in relation with his government . . . it was not dealing with children."[83]

In early March 1955, Dulles flew to Taiwan and met with Chiang Kai-shek. He started by apologizing for any confusion about the status of Jinmen and Mazu and acknowledged that he "was prepared to take full responsibility." However, Dulles went on, "the matter had developed in the US in such a way that the authority to use the Armed Forces of the US outside of the treaty area had to be left to the future judgment of the president of the US and that therefore there could not be any actual present commitment" to Jinmen and Mazu.[84] From this point on, the policy of "strategic ambiguity" was to be applied to the status of the offshore islands.

This policy had been chosen with not just China in mind: there was a very real threat that Washington dynamics could expand a small conflict in the Taiwan Strait into a world war. For example, Senator Karl E. Mundt (R-SD), who had been as a congressman a longtime member of the House Foreign Affairs Committee, urged Eisenhower on May 12, 1955, that if a Communist attack appeared "beyond the capacity of the Nationalists to repel" the United States should use its "air and sea power" against the PRC itself; "in 72 hours we could bring great destruction on the Mainland." After this "fierce attack," the senator suggested, leaflets could be dropped over the mainland "stating [that] the war—started by Communists—was over unless they attacked again."[85] It was just this kind of no-win scenario that Eisenhower was actively seeking to avoid.

Conclusions

In a show of force, the PRC initiated the first Taiwan Strait crisis during early September 1954. Scholars have argued that the PRC's primary motivation was to test "American resolve."[86] On April 26, 1955, William C. Bullitt, former ambassador to the USSR, supplied evidence for this view, writing to Eisenhower: "From Japan to Germany, the most serious men—even our devoted friends—have begun to doubt that we will have the intelligence and moral character to resist the Communist threat *in time.* The drift away from us is world-wide. If you defend Matsu and Quemoy, you will stop the rot in our world position. If you let them go, you will accelerate our decline and fall."[87]

Communist forces overran the Yijiang Islands and attacked the Dachen Islands. Washington repeatedly urged Taipei to withdraw from all the offshore islands, but Chiang Kai-shek refused. Not only would complete abandonment incur a loss of face for Chiang, but so long as the Nationalists held these islands right off China's coast the Communists could not proclaim total victory, that they had seized the entire mainland. In addition, Chiang argued, Chinese history showed that the Nationalists could someday invade the mainland from their offshore bases, using the disputed islands as stepping-stones to retake control of China.

From a strategic viewpoint, the PRC's attacks backfired. In December 1954, Taiwan and the United States signed a mutual-security pact affirming that the United States would defend Taiwan and the Penghu Islands. In a secret, supplementary pact, Chiang Kai-shek promised that no offensive operations were to take place against the mainland "without American agreement." Recounting this discussion four years later, Dulles was to recall that Chiang "had maintained that limitation and honourably maintained it."[88] Chiang had also agreed to evacuate, with American assistance, the Dachen Islands. Finally, the Formosa Resolution left it intentionally unclear which of the offshore islands Washington would choose to defend and which to evacuate. This ambiguity allowed tension in the Taiwan Strait to die down, at least temporarily.

Notes

1. Breakfast Discussion with Secretary of State John Foster Dulles, July 22, 1955, Top Secret, PREM 11/879, TNA.

2. U.K. High Commissioner in India, Report of Meeting with Prime Minister Nehru, November 10, 1954, Secret, FO 371/110238, TNA.

3. "The Struggle for the Coastal Islands of China," *ONI Review Supplement* (December 1953), pp. I–IX.

4. "Memorandum by the Attorney General on Congressional Attitude to Formosa Defense."

5. Conference with the President, May 22, 1954, p. 1, Dulles, J. F., WHM, box 1, Meetings 1954 (3), DDEPL.

6. "Preliminary Report of Van Fleet Mission to the Far East (Japan, Korea, Okinawa and Formosa)," June 3, 1954, Top Secret, WH Office, OSANSA, Special Assistant Series, President Subseries, box 2, President's Papers, 1954 (15), DDEPL.

7. Ibid.

8. Acting Secretary of Defense Robert B. Anderson, message to President Eisenhower, September 3, 1954, Top Secret, DH 4, Dulles, Sept. 54 (2), DDEPL.

9. General Smith, telephone call to President Eisenhower, September 6, 1954, DDE Diary Series 7, Phone Calls, June–Dec. 1954 (2), DDEPL.

10. Acting Secretary of Defense Robert B. Anderson, telephone call to President Eisenhower, September 4, 1954, DDE Diary Series 7, Phone Calls, June–Dec. 1954 (2), DDEPL.

11. Central Intelligence Agency, "The Chinese Offshore Islands," September 8, 1954, 33 pages, Top Secret, DDE AWF, Int. Series 9, Formosa (1), DDEPL.

12. Ibid.

13. NSC Talking Paper, September 12, 1954, Top Secret, JFD Papers, WHM 8, Gen. For. Pol. (4), DDEPL [emphasis original].

14. Ibid.

15. Robert R. Bowie, Department of State, Policy Planning Staff, memorandum for the Secretary, March 28, 1955, Top Secret, Dulles, J. F., WHM, box 2, Offshore April–May 1955 (2), DDEPL.

16. Memorandum for the Secretary of State, April 5, 1955, p. 1, Top Secret, Dulles, J. F., WHM, box 2, Offshore April–May 1955 (2), DDEPL.

17. Dulles, Draft "Formosa" Paper to Eisenhower, pp. 12–13.

18. NSC Talking Paper.

19. Memo, May 16, 1954, Top Secret, JFD Papers, WHM 8, Gen. For. Pol. (4), DDEPL.

20. Acting Secretary of State Dulles, telegram to the White House, September 9, 1954, Top Secret, DH 4, Dulles Sept. 54 (1), DDEPL.

21. NSC Talking Paper.

22. Ibid.

23. Bedell Smith, telephone call to President Eisenhower, September 29, 1954, DDE Diary Series 7, Phone Calls, June–Dec. 1954 (2), DDEPL.

24. Eisenhower and Dulles, phone conversation, October 5, 1954, DH 4, Dulles, J. F., Oct. 54, DDEPL.

25. John Foster Dulles, message to Eisenhower, Top Secret–Eyes Only, October 12, 1954, DH 4, Dulles, J. F., Oct. 1954, DDEPL.

26. Memorandum for the President, December 18, 1954, DH 4, Dulles, John Foster, Dec. 1954 (2), DDEPL, quoting NSC 5429/4.

27. Mr. Trevelyan's Conversation with Chou-en-Lai on 28 January 1955, Top Secret, DDE AWF, Int. Series 10, Formosa Area, U.S. Mil. Ops. (1), DDEPL.

28. Eisenhower, telegram to John Foster Dulles, February 21, 1955, Top Secret, DH 4, Dulles, J. F., Feb. 1955 (1), DDEPL; responding to Dulles, telegram to Eisenhower, February 21, 1955, Top Secret, DH 4, Dulles, J. F., Feb. 1955 (1), DDEPL.

29. John Foster Dulles, telegrams to President Eisenhower, February 25, 1955, Top Secret, DH 4, Dulles, John Foster, Feb. 1955 (1), DDEPL [emphasis original].

30. Dulles discussions with Chiang Kai-shek, March 4, 1955, Top Secret, DDE AWF, Int. Series 10, Formosa (China), 1952–57 (4), DDEPL.

31. Summary of Remarks of . . . Dulles, March 18, 1955, p. 2.

32. Memorandum of Meeting with the Senators, April 27, 1955, April 28, 1955, Top Secret, Dulles, John F., Gen. Cor. box 1, Memos J–K (2), DDEPL.

33. John Foster Dulles, telegram from Vienna to the President, May 14, 1955, Top Secret, DH 5, JFD May 55, DDEPL.

34. Ambassador Henry Cabot Lodge, memorandum to John Foster Dulles, May 4, 1955, Secret, Dulles, J. F., Tel. Conv., box 4, May–Aug. 1955 (8), DDEPL.

35. Telephone Call from Allen Dulles, July 1, 1955, Dulles, J. F., Tel. Conv., box 4, May–Aug. 1955 (4), DDEPL.

36. Memorandum Citing Mr. Krishna Menon's Discussions with Zhou Enlai, June 9, 1955, pp. 2–3, Secret and Personal, Dulles, J. F., WHM, box 3, Meet. Pres. 1955 (4), DDEPL.

37. Notes on Eisenhower Discussion with Menon, DDE Papers 6, ACW Diary July 1955 (3), DDEPL [emphasis original].

38. Memorandum for the President: Americans Imprisoned or Detained in Communist China, July 18, 1955, DH 5, JFD July 1955, DDEPL.

39. Call from Allen Dulles, July 1, 1955.

40. Chiang Kai-shek to President Eisenhower, June 7, 1953, DH 1, Dulles June 53 (1), DDEPL.

41. CTF 72 Operation Order No. 2-A-53, May 1, 1954.

42. Cited in Harold Macmillan to John Foster Dulles, September 5, 1958, Top Secret, CAB 21/3272, TNA.

43. U.K. Ambassador, Washington, DC, to Foreign Office, January 29, 1955, Secret, PREM 11/867, TNA.

44. George W. Anderson Jr., *Reminiscences of Admiral George W. Anderson, Jr.,* Oral History 42, Naval Institute Oral History Program, Annapolis, MD [hereafter NIOHP], p. 272.

45. Preliminary Draft of Possible Statement of Position for Communication to the Republic of China, April 7, 1955, p. 9.

46. "Pressure and a Pact," *Newsweek,* December 13, 1954.

47. Hungdah Chiu, *China and the Taiwan Issue* (New York: Frederick A. Praeger, 1979), p. 160.

48. Report on U.S. Government Policies in Relation to China, n.d., p. 5.

49. "Commitments and Problems of the United States to the Republic of China," p. 11.

50. Telephone Call from Mr. Hagerty, December 22, 1954, Dulles, J. F., Tel. Conv., box 10, Nov.–Feb. 1955 (2), DDEPL.

51. Message for Transmittal to the Prime Minister, February 19, 1955, Top Secret, DDE Diary Series 9, DDE Diary, February 1955 (1), DDEPL [emphasis original].

52. Dulles discussions with Chiang Kai-shek, March 4, 1955.

53. President Eisenhower, letter to Prime Minister Churchill, February 19, 1955, Top Secret, PREM 11/879, TNA [emphasis original].

54. Central Intelligence Agency, "The Chinese Offshore Islands."

55. Samuel B. Frankel, *The Reminiscences of Rear Admiral Samuel B. Frankel,* Oral History 325, NIOHP, pp. 359–60.

56. JFDP, July 15, 1953, reel 204/205, 89555, Princeton Univ.

57. Ibid.

58. U.K. High Commissioner in India, telegram to Commonwealth Relations Office, November 10, 1954, Secret, FO 371/110238, TNA.

59. Telephone conversation, January 18, 1955, DDE Diary Series 9, Phone Calls, Jan.–July 1955 (3), DDEPL.

60. "Aerial Tactics of the Chinese Communists in Naval Operations," *ONI Review* (February 1955), pp. 85–87.

61. JFDP, January 20, 1955, reel 210/211, 92599, Princeton Univ.

62. U.K. Ambassador, Washington, DC, to Foreign Office, January 19, 1955, Secret, PREM 11/867, TNA.

63. Memorandum of Luncheon Conversation with the President, Personal and Private, January 19, 1955, Top Secret, Dulles, J. F., WHM, box 3, Meet. Pres. 1955 (7), DDEPL.

64. Meeting in the President's Office, January 21, 1955, Top Secret, DDE Papers 4, ACW Diary, January 1955 (2), DDEPL.

65. Memorandum for the Record, January 29, 1955, DDE Diary Series 9, DDE Diary, January 1955 (1), DDEPL.

66. Telegram to CinCPac Pearl Harbor, January 29, 1955, DDE AWF, Int. Series 9, Formosa (2), DDEPL.

67. Memorandum for the Record, January 31, 1955, DDE Diary Series 9, DDE Diary, January 1955 (1); DDE Ad files 29, Radford, Admiral Arthur W. (2), DDEPL.

68. CinCPacFlt Op-Plan 51-Z-55, January 27, 1955, Top Secret, DDE AWF, Int. Series 10, Formosa Area, U.S. Mil. Ops. (3), DDEPL.

69. U.K. Embassy, Washington, DC, to Foreign Office, February 6, 1955, PREM 11/867, TNA.

70. Telegram from U.K. High Commissioner in Australia, telegram to Commonwealth Relations Office, April 16, 1955, Top Secret, DEFE 13/288, TNA.

71. Dulles, telegrams to Eisenhower, February 25, 1955.

72. Notes Dictated by the President Regarding His Conversation with Sir Anthony Eden, July 19, 1955.

73. Memorandum of Conference with the President, December 20, 1954, DDE Papers 3, ACW Diary, December 1954 (2), DDEPL.

74. Meeting of Secretary with Congressional Leaders, January 20, 1955, p. 1, Personal and Private, Secret, Dulles, J. F., box 2, WHM 1955, FS (3), DDEPL.

75. Ibid., p. 4.

76. Ibid., p. 5.

77. Dulles and Eisenhower Telephone Conversation, January 22, 1955, DDE Diary Series 9, Phone Calls, Jan.–July 1955 (3), DDEPL.

78. President Eisenhower, letter to Prime Minister Churchill, January 25, 1955, Top Secret, DDE Diary Series 9, DDE Diary, January 1955 (1), DDEPL.

79. Formosan Straits, Substance of a Message Dated January 28 from Mr. Trevelyan in Peking, January 28, 1955, Secret, DH 4, Dulles, J. F., January 1955, DDEPL.

80. The full name of the Formosa Resolution is the "U.S. Congressional Authorization for the President to Employ the Armed Forces of the United States to Protect Formosa, the Pescadores, and Related Positions and Territories of That Area."

81. Ambassador Rankin to Secretary of State, January 29, 1955, DDE AWF, Int. Series 10, Formosa Area, U.S. Mil. Ops. (3), DDEPL.

82. Washington, DC, telegram to Taipei, January 29, 1955, Top Secret, DDE AWF, Int. Series 10, Formosa Area, U.S. Mil. Ops. (3), DDEPL.

83. Taipei, telegram, January 30, 1955, Top Secret, DDE AWF, Int. Series 10, Formosa Area, U.S. Mil. Ops. (3), DDEPL.

84. Dulles discussions with Chiang Kai-shek, March 4, 1955.

85. Memorandum to the president regarding Senator Mundt's letter, May 12, 1955, DDE Papers 5, ACW Diary May 1955 (5), DDEPL.

86. Thomas J. Christensen, *Useful Adversaries: Grand Strategy, Domestic Mobilization, and Sino-American Conflict, 1947–1958* (Princeton, NJ: Princeton Univ. Press, 1996), p. 195.

87. William C. Bullitt, personal letter to President Eisenhower, Personal, April 26, 1955, DDE Diary Series 10, DDE Diary April 1955 (1), DDEPL.

88. Foreign Office, telegram to U.K. Embassy, Washington, DC, October 22, 1958, PREM 11/3738, TNA.

CHAPTER FOUR

The Growing Militarization of the Offshore Islands

A boat can't always sail with the wind; an army can't always win battles.
(风无常顺, 兵无常胜.)

As discussed above, Washington used economic and military tools. A strategic embargo was adopted by the U.S. government to put additional pressure on China. On February 10 and March 22, 1949, meetings were held between American and British officials regarding putting controls on foreign exports to China. In November 1949, a report commissioned by Dean Acheson, the American secretary of state, declared, "We have been reluctant thus far to impose unilaterally new controls over exports to China because of the possibility that such action would handicap our negotiations with the British." After it had been pointed out to them "that such controls would represent the most important single instrument available for use vis-à-vis the Chinese Communists" the British finally agreed to control "1A list exports to China," provided that the other major powers also agreed to do so; to put limits on petroleum sales; and, as mentioned above, to "observe and to exchange information with the United States . . . on the movement of 1B goods to China with a view to joint consultation regarding corrective measures if it appeared that the flow was excessive or injurious to our common interests."[1]

Throughout the early 1950s, the British and the Americans closely watched the development of the Nationalist navy and the People's Liberation Army Navy (PLAN). In November 1949, the British listed the Nationalist naval assets as "7 destroyers, 21 destroyer escorts and 22 gunboats," backed by a 300,000-man army and a 230-plane air force composed of "145 Fighter bombers, 60 medium bombers and 25 Heavy bombers." From these numbers the British Chiefs of Staff concluded that a "direct invasion of the island [Taiwan] is, however, improbable for several months provided the Nationalist Naval and Air Forces remain loyal, since the Communists have inadequate naval and air forces to support the passage of troops and transports."[2]

A prime U.S. goal was to keep the Nationalist navy roughly equivalent to the PLAN but not greatly superior, so as to avoid the possibility of a major Nationalist attack on the mainland. In July 1951, the PLAN was still being described by the ONI as a "pick-up" fleet, the "youngest and weakest component of Red China's armed forces." In particular, it was "still handicapped by the lack of equipment and trained personnel." But the Communist ships' firepower—even discounting the still-unserviceable former British cruiser *Chongqing*—comprised thirty-nine guns over three inches in caliber, compared with the Nationalists' twelve. Furthermore, the majority of China's largest and best ships were based at the East China Military District Naval Headquarters in Shanghai, which faced Taiwan.[3] Politically too there was an obstacle to carrying out this plan of limited support, one arising from the generally poor relations between Eisenhower and Chiang.

Dwight D. Eisenhower and Chiang Kai-shek

U.S.-Taiwanese relations were not always easy, and they were often quite tense. During this entire period, the Nationalists conducted a blockade against mainland China. The Taiwanese government was eager to solidify relations with the United States. But one CIA report even warned that "aware of US interest in that island, they [the Nationalists] will present themselves as a means and perhaps the sole means of preventing its communization, and will offer various inducements and assurances in return for US aid and US moral support for a regional Chinese regime."[4]

Chiang would not stand for being treated as an inferior. Many leaders in America and in Europe, however, considered Chiang Kai-shek to be acting like a child—which his injured telegram of January 1955 (chapter 3) suggests that he suspected. They doubted his judgment, his words, and his goals. At a NATO meeting in May 1955, when future secretary general Paul-Henri Spaak criticized Chiang, Dulles defended him by "jokingly" replying that "while greater perfection could doubtless be found in Europe, it was not readily found in Asia."[5] One American officer criticized the Nationalists' "highly centralized control," and was convinced that it was the "awe in which they hold their President [that] conspired to keep him ill informed of the military situation and stifles initiative and daring of his subordinates all the way down."[6] At the height of the 1958 crisis, Eisenhower told Gen. Nathan F. Twining, USAF, then chairman of the Joint Chiefs of Staff, that "something must be done to make Chiang more flexible in his approach."[7]

Some critics of Chiang were even less generous. In January 1951 a Mr. Smith bluntly told Truman "that an effort be made to find a leader to replace Chiang Kai-shek."[8] Although many supported this idea, the American policy had been firmly established since December 31, 1948, when Ambassador Stuart told a Chinese political leader trying to oust Chiang that assisting him would be tantamount to intervention: "The question of bringing about Chiang's resignation or setting up a successor government is, in our opinion, a matter for Chinese decision and the US government could, under no circumstances, place itself in the position of dictating or suggesting the resignation of the head of a friendly government."[9]

But just because leaders in Washington did not agree to overthrow Chiang did not mean they liked him. Gen. Alfred Gruenther (Supreme Allied Commander Europe) referred to Chiang in early February 1955 as a "palooka."[10] The deputy director of the U.S. Information Agency (USIA), Abbott Washburn, told Eisenhower that the "8 million Taiwanese would be happier without either Chiang or Mao."[11] As to Chiang's determination to defend the offshore islands at all costs, Dulles complained, "He is gambling his whole position in Taiwan and his future as a useful agent in helping to drive communism from China against a local and possibly temporary success in a precarious defense of two island groups which are militarily weak."[12]

Negotiating with Chiang was particularly challenging and tedious. He often threw temper tantrums and made ultimatums to get his way; he had done so, for example, during World War II with the American commander of the China Burma India Theater, Gen. Joseph Stilwell (who called him "Peanut"). On February 21, 1955, Eisenhower telegraphed Dulles a list of suggestions with which to approach Chiang but added, "As you and I have agreed, any approach to Chiang along this line would have to be so skillfully conducted as to make him ostensibly the originator of the idea."[13] Later, the point came up again: "The crux of the negotiations must be that 'Chiang must sell himself on the validity and value of the suggested program.'"[14] If successful, the "worldwide political advantages of such an arrangement would be incalculable," but to be avoided at any cost was a "public belief that the alterations came about through American intervention or coercion."[15]

In fact, Chiang's frequent ultimatums culminated in a threat, if the Americans refused to help defend the islands, to expose his own Nationalist forces to destruction attempting to do so alone, just to spite the United States. It was truly a "cut off your nose to spite your face" threat. Eisenhower simply would not buy it. The U.S. government, after all, had put its own forces at risk assisting the Nationalists to evacuate the Dachen Islands. He had no intention of putting even more American troops at risk by militarizing the offshore islands. On February 1, 1955, Eisenhower told Radford that it was out of the question: "I have no intention of putting American foot soldiers on Quemoy. A division of soldiers would not make any difference."[16]

Eisenhower was firm that in any fight with the Communists over these islands, it would be up to Chiang to supply all of the troops:

> It is the Chinese Nationalists, not the Americans, who are contenders for the support of the Chinese people. It is the discipline, loyalty, and will to fight of the Chinese Nationalists which has been consistently disparaged by Communist propaganda. A battle which they won, on their own, or even a battle which they lost under conditions which would reflect high honor on the vanquished in the

face of overwhelming odds, would be far more advantageous to the Chinese Nationalists than a defense which could be sustained only by United States might, particularly if that was expressed in atomic terms.[17]

Eisenhower flatly dismissed Chiang's claim that the offshore islands were absolutely critical to Taiwan's defense, telling Dulles, "If this is so, his own headquarters *should* be on the offshore islands," so as to show his determination to hold them.[18]

On February 1, 1955, Eisenhower wrote a six-page, top-secret letter to General Gruenther at his NATO headquarters. One of its goals was to explain why the Formosa Resolution's "wording, as to areas outside Formosa and the Pescadores, is vague." Eisenhower had to balance the expectations of four parties: Taiwan, China, Europe, and America. In military terms as well as political, the defense of Taiwan and the Penghus, only about thirty miles west of Taiwan, was very different from defending the offshore islands, close to the mainland:

> Any military man can easily make clear distinction between the defense of Formosa and the defense of the so-called offshore islands. Not only are two different military problems presented, but in the one case we are talking about territories the control of which has passed from nation to nation through the years—and in the other case, about territories that have always been a part of the Chinese mainland both politically and, in effect, geographically. So the political differences are almost as plain as the military differences when we talk about the defense of these two territories.

Considering even military factors alone, then "we would permit *no* advance by the Communists beyond the offshore islands, but . . . in any struggle involving only the territory of those islands, we would see no reason for American intervention."[19] While Europeans and Americans would be pleased with this solution should it become public, Eisenhower foresaw, the PRC and the ROC would both be infuriated, since the former wanted to retake Taiwan and the latter wanted to retake mainland China. However, if the U.S. government stated clearly that it planned to defend Jinmen and Mazu, the Nationalists would be pleased, the PRC would be even more infuriated, and the Europeans would be frightened. There really wasn't any single solution that satisfied all parties.

Aside from being drawn into China's civil war, the president could see that defending these offshore islands would require huge military resources:

> By announcing this as a policy we would be compelled to maintain in the area, at great cost, forces that could *assure* the defense of islands that are almost within wading distance of the mainland. This defensive problem could be extremely difficult over the long term, and I think that the world in general, including some of our friends, would believe us unreasonable and practically goading the Chinese Communists into a fight. We could get badly tied down by any such inflexible public attitude.

The only possible solution was to be ambivalent about the status of the offshore islands, thus allowing the president the power to make a distinction "between an attack that has *only* as its objective the capture of an offshore island and one that is *primarily a preliminary movement to an all-out attack on Formosa*." The whole problem was a "can of worms," Eisenhower admitted, but "whatever is now to happen, I know that nothing could be worse than global war."[20]

On February 9, 1955, the Senate accepted the U.S.-Taiwan pact, and it was ratified on March 3, 1955. This rapid ratification reflected acceptance of Dulles's argument that in part the PRC's offshore attacks were an attempt to "scare us out of ratifying the treaty." He predicted that with "ratification an accepted fact, it might take some of the heat out of the situation," and so it proved.[21] Even before, Dulles told Eisenhower on February 16, Chiang had been convinced to accept four important points: to agree to the Formosa treaty; to "agree to use of no [American] equipment or people we have trained outside Formosa & the Pescadores"; "to acquiesce on the UN move for cease-fire"; and "to evacuate the Tachens."[22]

Dulles, like Eisenhower, was concerned that a major war over the offshore islands could lead to disaster: "If conflict in that region should spread to global proportions, we would be entering a life and death struggle under very great handicaps."[23] But he also warned that "we cannot at this time squeeze any more out of him [Chiang]." A carrot, he thought, might be more effective than a stick—perhaps a promise that American aid would be increased. Chiang was to be told that "we're working like dogs in an attempt to keep 75 F-86's up to snuff, & keep training program going along so that they always have 75 pilots to man them."[24] On February 19, 1955, Eisenhower told Admiral Radford that "he would like to see Chiang make proposals on his own initiative which would ease the situation concerning the off-shore islands and improve the U.S. security position in the Formosa area." Only then would Eisenhower "be inclined to provide reserves, and would seriously consider maintaining U.S. forces of the order of a battalion of Marines and a squadron of F-86's on the island."[25]

Two days later, on February 21, 1955, Eisenhower elaborated to Dulles: "You will recall also that you and I talked about certain other things we might do to convince him [Chiang] that his best course of action lay in solidifying his union with us, so as to insure preservation of Formosa and the Pescadores without risking too much of his force in the forward positions. I refer to our readiness to speed up his air development and possibly strengthen both naval and air units above presently contemplated levels."[26] But convincing Chiang to reduce troop levels on, or even abandon, the offshore islands would prove a major challenge.

Discussions on Abandoning the Offshore Islands

The true problem for American military and political leaders in the first Taiwan Strait crisis was to convince Chiang Kai-shek to settle the dispute peacefully, preferably by abandoning the offshore islands completely, and without undermining the international perception of U.S. support for Taiwan. If the PRC took the islands by force, Chiang would lose prestige.[27] As Eisenhower reminded Churchill on January 25, 1955, the United States had to be concerned with the "solidarity of the Island Barrier in the Western Pacific"; for Washington to desert Taipei might "risk a collapse of Asiatic resistance to the Communists."[28] That, in turn, would weaken the global U.S. containment policy against communism. A month later, he told Churchill that the United States "does not have decisive power in respect of the offshore islands" and that since it "must not lose Chiang's army and we must maintain its strength, efficiency and morale," attempts to coerce him might easily backfire.[29]

There were potent historical arguments that the offshore islands might serve as stepping-stones across the Taiwan Strait. But in March 1955, President Eisenhower, perhaps the world's greatest expert at that time on amphibious landings, disputed that line of reasoning in a personal letter to a longtime friend, Lew Douglas: "Those particular off-shore islands have no value whatsoever as stepping stones in any possible future attempt of Chiang to go back to the mainland. You do not conduct an amphibious operation against a hostile coast by first going onto some tiny islands and then launching a second amphibious attack."[30] In any case, Eisenhower pointed out to Lew (whose side of the conversation is unavailable), "you do not tell me *what to do if we lose Formosa.*" While the offshore islands were useless for attack (in either direction), they still had military value: "Defensively they practically block almost any Communist attempt to use the two available harbors immediately west of Formosa for the initiating of amphibious operations." So, "What I am asking you is this: If you became convinced that the capture of these two places by International Communism would inevitably result in the later loss of Formosa to the free world, what would you do? Beyond question the opinion in Southeast Asia is that the loss of Formosa would be catastrophic; the Philippines and Indonesia would rapidly be lost to us."[31]

Two weeks later, Eisenhower wrote Lew of the 750,000 Nationalist Chinese who wanted to return to China, to whom he argued the offshore islands "rightly or wrongly" symbolize doing so: "It happens that I believe they [the islands] should *not* mean this. They have no value as stepping stones to the mainland. (I repeat that they do have some defensive value for Formosa.) The problem becomes how do we induce the seven hundred and fifty thousand Chinese in Formosa to maintain their morale, their spirit, their hopes, when we ask them to give up two areas that have been for them the symbol of their deepest aspirations." If Washington did not keep Nationalist morale high, "Internal disillusionment, despair and the unquenchable hope of going back home will establish the Communist regime" on Taiwan.[32]

If the United States had few choices in the present situation, Dulles pointed out, the range of possible reactions to a PRC attack in the Taiwan Strait was huge: "It is important to recognize the intrinsically great strategic strength of the U.S. position in the Formosan crisis and the great freedom of choice we possess." For example, Washington could adopt a business-as-usual approach, or exercise deterrence, or "egg on" the PRC to attack at a place and time of America's choosing, or try to reach some kind of diplomatic settlement. If deterrence were adopted, the implied threat could be widened to the northward by abrogating the Korean armistice so as to create a potential second front. Or to the south: increased fighting in Indochina (that is, by the South Vietnamese forces of then–prime minister Ngo Dinh Diem against the communist Vietcong) could divert China's attention away from Taiwan. Short of actual war, then, a wide selection of military avenues could be used to "persuade the Chinese Communists that we not only have the means to stop them but we will use these means to defeat them if they challenge us."[33]

However, as matters then stood, if Chiang withdrew from the offshore islands of his own volition, he could not possibly "lose face." Neither would he if the PRC took the islands by force: "such loss would occur only after the defending forces had exacted a fearful toll from the attackers, and Chiang's prestige and standing in Southeast Asia would be increased rather than decreased as a result of a gallant, prolonged and bitter defense conducted under these circumstances." Nationalist losses in such a defense would be "inconsequential both in personnel and in material" terms, but the PRC's "should be very great indeed."[34]

By the end of February 1955, it looked more and more likely that war would break out over the Taiwan Strait. Dulles told Eden on February 25 that "whereas up to [a] few weeks ago we had believed Chinese Communists were not seriously intending [to] take Taiwan by force, we now believe they intend to do so. So in fact we are in a battle for Taiwan." To cede Taiwan to the Communists could negatively impact all of Asia: "Further retreat could swing Asia. Trends in Japan are already disturbing. Further retreat or loss of Formosa would convince Japan [that] communism [was] wave of future. Consequent effect on Okinawa and other parts of Asia obvious. Overseas Chinese would turn to Peking."[35] Even the loss of just the offshore islands, if attributable to American "timidity," might "create doubts in Japan and concern among our friends in Asia, particularly in Thailand, the Philippines and Korea."[36] Eisenhower could not cut Chiang loose, therefore; there was little choice but to support him.

Eisenhower Offers Greater Aid to Chiang

Even while attempting to reduce military tensions in the Taiwan Strait, Eisenhower had to be aware of how a Nationalist retreat might look to the rest of East Asia. The perceptions of overseas Chinese were critical, for example, particularly in the Philippines, Indonesia, Malaya, and Hong Kong. If the Nationalist alternative to the PRC disappeared, Eisenhower told Churchill, "these émigré Chinese will certainly deem themselves

subjects of the Chinese Communist Government and they will quickly add to the difficulties of their adopted countries." Eisenhower then warned that the overseas Chinese
influence might become so strong that no amount of foreign pressure could "prevent these
regions from going completely Communist. Do not such possibilities concern you?"[37]

During spring 1955, Eisenhower tried on several occasions to convince Chiang to give
up Jinmen and Mazu Islands, but Chiang refused.[38] In early April, Eisenhower offered
Chiang "up to a division" of U.S. troops plus "an air wing" if "Quemoy and Matsu were
made outposts rather than strongholds," which would lessen their importance, and so
vulnerability, as "symbols of prestige."[39] Eisenhower presented five goals:

(1). To regard the offshore islands as outposts and consequently to be garrisoned in accordance with
the requirements of outpost positions. This involves vigilant reconnaissance and a maximum
of protective works and with properly sited automatic weapons and light artillery, together with
effective obstacles, defensive mine systems, and so on. All this should be reinforced by adequate
stores of ammunition, of food and medical supplies, all thoroughly protected and available to
the garrison as needed. Excess personnel (except such civilians as might decline to leave) should
be removed from the islands.

(2). The Nationalist forces on Formosa should assist these garrisons by aerial and sea reconnaissance and fighting support. Plans for defense should be fully coordinated between the forward
units and the mobile elements in Formosa.

(3). Adequate plans should be made for determined and persistent defense, and evacuation should
take place (if this finally becomes necessary) only after defensive forces had inflicted upon the
attackers heavy and bloody losses.

(4). The process of concentrating, equipping and training of troops on Formosa itself should be
expedited. The United States could and would help in this process so as to give to Chiang the
greatest possible strength in support of his outpost troops on Quemoy and the Matsus, and in
preparing and sustaining the bulk of his forces as a weapon of opportunity, ready to take advantage of any political, military or economic circumstance on the mainland that would give to an
invasion a reasonable chance of success.

(5). To protect the prestige of Chiang and the morale of his forces, any alteration in military and
political planning should obviously be developed under his leadership; above all, there must
be no basis for public belief that the alterations came about through American intervention or
coercion.[40]

Discussions of a retreat from Jinmen and Mazu were under way in Washington. In
return for Chiang's acquiescence, it was proposed, Eisenhower would promise to create
a joint U.S.-Taiwan "defense zone" from Shantou to Wenzhou "in which the movement
of all seaborne traffic of a contraband or war-making character would be interdicted."
In particular, the U.S. Navy would lay minefields that "would force coastwise junk
traffic to come out where it also could be intercepted and controlled."[41] According to an
early draft, this plan would serve a triple purpose: "It would replace Quemoy and Matsu
as defensive blocks to the staging of a seaborne attack on Formosa from Amoy and
Foochow harbors; it would materially curtail the present seaborne movement of POL
[petroleum, oil, and lubricants] and like supplies into the Fukian airfield area—an area

which cannot be easily supplied by land; it would demonstrate that the United States is prepared to take strong measures in the defense of Formosa."[42]

In April 1955 Admiral Radford, then newly appointed as chairman of the JCS, and Walter Robertson, Assistant Secretary of State for Far Eastern Affairs, flew to Taiwan to discuss this proposal with Chiang. They promised that the United States would supply better naval equipment to patrol the Nationalist blockade if Chiang would give up the disputed offshore islands. Chiang asked Robertson to confirm that "President Eisenhower's proposal was to give up Quemoy-Matsu and substitute there for interdiction [of a] limited area [of the] China coast." Robertson did so, "emphasizing that under present circumstances attack on communist build-up would involve us striking first blow whereas interdiction would put communists into position of striking first blow."[43]

According to one recollection of this meeting, Chiang asked for a minute to consider the president's offer and left the room, supposedly to pray in the garden. He returned and reportedly said, "I just want to tell you that I have prayerfully concluded that I cannot accept such a proposal because I do not have the faith in your government to sustain it."[44] The transcripts have Chiang declaring that he would "defend Quemoy-Matsu with or without US help."[45] Primarily, Chiang vetoed the proposal for reasons of "face": relinquishing control over additional offshore islands would make his government look weak. Later explaining the situation to Anthony Eden, Eisenhower said the offshore islands "should be considered as an outpost, but Chiang has said that abandonment of the islands would result in loss of face and of any hold over Chinese in Malaya and elsewhere in the Far East."[46] British recognition of the PRC was also a factor for Chiang: once he gave up Jinmen and Mazu, the United States might soon halt any "effective shipping interdiction scheme in the face of strong and inevitable opposition by the British and others."[47]

In any case, Chiang's refusal left the U.S. government with few options: it had either to support Chiang on his terms or risk losing him as a dependable ally in the midst of the Cold War. Eisenhower's sense of "the situation was simply that if we tried to press Chiang too hard to give up the islands, Formosa might be lost and the whole position in the Far East might crumble."[48]

Eisenhower was not surprised when he heard that Generalissimo (or "Gimo") Chiang had not agreed to Washington's proposal:

> I had hoped that the Gimo himself might have seen the wisdom of trimming the garrison on the offshore islands down to the leanest fighting weight possible, organizing them highly, and in the meantime making the necessary public statements that would clearly set forth his determination to fight for the islands' positions, but *not* to make them the sine qua non of the ChiNats' existence. I had thought also that while he was doing this, if he could be assured of our reinforcing Formosa with air, some marines and logistics, that he would have been in better position both politically and militarily than he now is. Certainly this would have been better for us.

Eisenhower admitted, however, that Chiang's response was "not only what I predicted but what I think I would have made had I been in his place." This was partly due to the complexity of the situation; in any case, since "it is clear that as long as Radford and Robertson themselves could not grasp the concept [i.e., Eisenhower's viewpoint], we simply were not going to get anywhere."[49]

Back in Washington, Radford and Robertson on May 3 explained to the president that while Chiang would not agree to withdraw in return for the "imposition of some kind of interdiction of the Formosa Strait sea area," he did agree that the offshore islands should be defended only by Nationalist forces. To that end, Admiral Stump would visit Taiwan and "advise Chiang as to ways in which defense preparations in the offshore islands can be improved." President Eisenhower fully backed a general policy of beefing up defenses, and the State Department agreed to pass through the American ambassadors in the region the "President's assurances of our willingness to provide logistic support for the build-up of the offshore island defense."[50] The new support would include providing more "automatic weapons," constructing "undersea obstacles," "laying antipersonnel mine fields on the beaches," "increasing barbed wire defenses," and providing "hard rock drills" with which to expand the underground bunkers.[51]

In mid-1955, the combined intelligence staff of CinCPac and his subordinate Commander in Chief, Pacific Fleet produced an estimate of the offshore islands that was supposed to be relevant through mid-1956. It assessed that

> the greatest importance of the offshore islands at present is political and psychological. Any further gains in these islands by the CHICOMS will give the CHICOMS a further boost in morale and add to the string of Communist victories in Asia, i.e. mainland China, Korea, Indochina and the Tachens. Conversely, any further loss or yielding of these islands will be a serious blow to CHINAT morale and regarded by the remaining anti-Communist nations in Asia as a further disastrous retreat by the U.S., since at present, the U.S. is so closely identified with the Chinese Nationalists that any CHINAT reverse will be viewed as a U.S. loss.[52]

The Chinese communists' most probable course of action would be to keep trying to seize the "smaller, more weakly defended" offshore islands but not either Formosa or the Penghus, especially if "the U.S. is firmly committed to defense of these islands." Meanwhile, the "Chinese Nationalists will attempt to retain all of the territory they now hold, putting pressure on the U.S. for assistance. They will not willingly make any further withdrawals."[53] Retention of the islands was, as has been noted, particularly important to their morale.

The Problem of Nationalist Morale

Eisenhower had to be concerned with not only things military but also Nationalist morale and possible accusations against his and Truman's administrations of retreating in the face of aggression. As we have seen, the U.S. government felt obliged to support

Chiang Kai-shek's control over a number of offshore islands, even though, for example, in 1954 the Seventh Fleet had been unable to focus entirely on Taiwan. As Eisenhower assured Albert C. Wedemeyer (a retired general with strongly anticommunist and pro-Nationalist views), keeping up the Nationalist morale was critical: "if an all-out attack should ever develop, they are the ones who must do the land fighting."[54] American naval demonstrations, designed to warn the PRC not to attempt an invasion, were particularly important means of doing so (see chapter 5).

The issue was very important throughout the policy establishment. A report that examined only military factors angered Dulles, for instance: its "analysis neglected critical factor of effect on morale of loss of offshore islands. . . . Very considerable factor in situation now is possibility of deteriorating morale at Taiwan. Withdrawal from islands might have critical effect on ability [of] Chinese Nationalists [to] hold islands if morale disintegrated and groups there made deals with Communists."[55] If the PRC attacked the islands, Eisenhower was sure, U.S. failure to assist "would dismay the ChiNats, *whose morale and military efficiency are essential to the defense of Formosa—and the security of Formosa is essential to the best interests of the United States and the Western world.*" Further, to persuade Chiang now to give them up "*would result in a collapse of morale on Formosa and the loss to the free world of that bastion of strength.*" In fact, "the principal military reason for holding these two groups of islands is the estimated effect of their loss upon morale in Formosa."[56]

The president's last point was the crux of the entire question of defending the offshore islands or not—that is, it rested on Nationalist morale. The American embassy in Taipei had an important role to play in morale building. Throughout the 1950s there was well-founded concern that the morale of average people on Taiwan was weakening; after all, they were a small island nation facing alone an enormous continental power. As Ambassador Rankin explained to a British visitor in 1953, "the Chinese Nationalists on Formosa represented the only case where a sizeable element of a communist dominated country had escaped from behind the iron curtain and was conducting his affairs as an independent government." Taiwan's success or failure could have a dramatic effect on "the whole anti-communist problem[,] . . . in the ultimate show-down." Taiwan's "unique position" required "great vision and foresight in framing our future policy toward communism as a whole, and in particular, towards the situation in the Far East."[57]

The result was a delicate balancing act, keeping the Nationalists happy but not giving them too much leeway. In a February 10, 1955, top-secret letter for Churchill's eyes only, Eisenhower described the basic problem:

> To defend Formosa the United States has been engaged in a long and costly program of arming and sustaining the Nationalist troops on the island. Those troops, however, and Chiang himself, are not content, *now,* to accept irrevocably and permanently the status of "prisoners" on the island. They are held together by a conviction that some day they will go back to the mainland.

As a consequence, their attitude toward Quemoy and the Matsus, which they deem the stepping stones between the two hostile regions, is that the surrender of those islands would destroy the reason for the existence of the Nationalist forces on Formosa. This, then, would mean the almost immediate conversion of that asset into a deadly danger, because the Communists would immediately take it over.

The Formosa Resolution, as passed by the Congress, is our publicly stated position; the problem now is how to make it work. The morale of the Chinese Nationalists is important to us, so for the moment, and under existing conditions, we feel they must have certain assurances with respect to the offshore islands. But these must be less binding on us than the terms of the Chino-American Treaty, which was overwhelmingly passed yesterday by the Senate. We must remain ready, until some better solution can be found, to move promptly against any Communist force that is manifestly preparing to attack Formosa. And we must make a distinction—(this is a difficult one)—between an attack that has *only* as its objective the capture of an offshore island and one that is primarily a preliminary movement to an all-out attack on Formosa.

Whatever now is to happen, I know that nothing could be worse than global war [which must have been much on Ike's mind just then; he had used the same phrase to General Gruenther ten days before]. . . . I devoutly hope that history's inflexible yardstick will show that we have done everything in our power, and everything that is right, to prevent the awful catastrophe of another major war.[58]

As Eisenhower emphasized in a later letter to Churchill, in some ways the U.S. government had gained much less than it wanted in negotiations with Taipei, since it had been careful not to coerce Chiang Kai-shek lest it "undermine the morale and the loyalty of the non-Communist forces on Formosa."[59]

One major problem was that the Americans had to show that they supported Taiwan while not giving an open-ended promise to help defend Jinmen and Mazu. During June 1955, for example, Chiang Kai-shek informed Washington that he wanted to reinforce the offshore islands with an additional division but that he wanted first to learn what the American view was, since "should the public and military learn that we [Americans] oppose reinforcement of the island garrisons, they might deduce that we are thinking of urging another Tachen-like withdrawal."[60] Clearly, any discussion of troop movements had to take into account how they would impact morale within the Nationalist military.

Increased U.S. Military Aid to Taiwan

Both to strengthen the Nationalists' military position and to prop up their morale, the U.S. government increased military aid. In December 1954, Eisenhower told Foreign Minister Yeh that Washington would allocate an additional $106,225,000 to "(1) Train and maintain 63,000 additional officers and men 'at one time,' exclusive of current reserve training (30,000 every four months); (2) stockpile clothing and equipage for 341,700 men against an emergency; (3) fill 20,700 vacancies to bring nine army divisions up to full strength."[61] As Arleigh Burke saw it: "So we gave them lots of training and we gave them lots of [dock landing ships] and ships, boats, to supply the thing." With U.S. training and equipment the Nationalists duly "reinforced their garrisons and put their supplies in caves."[62]

The Nationalist navy depended on the U.S. Navy for equipment and for training in how to use it. Public Law 188, signed on August 5, 1953, authorized the United States to lend or give naval ships to "friendly nations in the Far Eastern area." To assist the Nationalists in defending the offshore islands, the JCS allocated ten patrol-type craft, two landing craft repair ships, and about a hundred small landing craft (LCMs and LCVPs).[63] The Department of Defense agreed to consider a Taiwanese request for four destroyers, six destroyer escorts, thirty landing ships, and six minesweepers;[64] however, the destroyers were questioned. One footnote in the CNO's (Carney's) memorandum to the JCS declared that "delivery of any additional destroyers . . . beyond the two now planned for [fiscal year] 1954, will be affected [sic] only upon the clearly demonstrated capability of the [Nationalist Government of the Republic of China] to man and operate them."[65]

On January 15, 1955, the NSC issued a top-secret "statement of policy" for the Taiwan Strait. An annex summarizing the direct military support and economic assistance being given to Taiwan in 1951–56 listed economic aid as $636,100,000, out of an overall total of almost $1.181 billion from 1950 to 1956.[66] The NSC specified six ways to support Taiwan's economic growth: provide technical and economic assistance, encourage Taiwan to join noncommunist Asian economic groups, encourage greater trade with noncommunist nations, urge Taiwan to make productive use of its own natural resources, help Taipei adopt better fiscal procedures, and stimulate both Chinese and private capital investment to develop Taiwan's economy. Finally, by using Taiwan as an example of the "soundness of U.S. policy," the United States could attempt "to convince other Free World countries . . . of the advisability of their adopting similar policies."[67]

Washington's most important goal was to preserve, with UN support, the "status quo" of the disputed offshore islands while providing the Nationalists sufficient military equipment and training to defend them. Direct American support for Taiwan's military was also important. The total U.S. military aid to the ROC between 1950 and 1956 was $1,181,100,000, spent to create a Taiwanese navy of three destroyers, six destroyer escorts, and eighty-two smaller combatant vessels, supported by an air force of eight and a third air wings. The goal was to build a force able to defend not only Taiwan and the Penghus but the offshore islands as well, and also assume responsibility for "conducting limited coastal patrols, anti-shipping, and commando operations." However, these military assets were, by design, insufficient either to "defend its [Taiwan's] territory successfully against full-scale Chinese Communist attack" or "to initiate large-scale amphibious operations against the mainland of China" without "U.S. air, naval and logistic support."[68]

By the mid-1950s the U.S. Navy had also given Taiwan two old diesel submarines. As Vice Adm. Philip Beshany, Commander, U.S. Taiwan Defense Command from 1972 to 1974, later recalled, the handpicked Nationalist crews were trained at the U.S. Naval

Submarine School, in New London, Connecticut. They were "obviously top caliber" and as students "absolutely excelled." The instructors later claimed that the Nationalist officers took to submarine operations "like ducks to water." Because the two submarines were meant only to help train the Nationalist navy in antisubmarine warfare, the United States did not provide torpedoes; still, Beshany thoughtfully concluded, "I often wonder if they didn't somehow get some torpedoes from some other country."[69]

New airplanes were also crucial to Taiwan's defense. Of the $132 million in total equipment deliveries between 1951 and 1954, fully two-thirds were in the form of "aircraft and aircraft spares," while another 18 percent were "bombs, rockets and ammunition." In 1954 alone, the United States shipped $48.3 million in equipment to Taiwan. By December 1954, 456 of the 657 aircraft promised to Taiwan had been delivered, 131 of them during that year. These included "seventy-two F-84G's, twenty-five F-86F's, sixteen T-33's and five RT-33's."[70] By December 1956, U.S. military assistance had created on Taiwan "21 regular infantry divisions, eight air wings, a small but efficient navy and marine corps, plus various special combat and support units."[71] Airpower was to play a particularly important role in future Taiwan Strait crises.

Conclusions

Supporting the Nationalists' military and morale was a major American goal. The delivery of high-tech equipment and training in its use proved invaluable. Between 1950 and 1969, American military aid to Taiwan reached $3.19 billion.[72] With equipment and training from the U.S. Navy, the Nationalist navy gradually improved. Eisenhower was careful, however, not to underestimate the PRC's military capabilities. As he told Churchill, "Because the ChiComs have no great fleet and cannot now attack across the seas, it is natural to underestimate their potential strength. . . . So I believe it critically important that we make a sober estimate of what we are up against."[73] In addition to giving Taiwan new offensive weapons, therefore, American advisers in Taiwan helped build airfields and provided better radar equipment, all to allow the Nationalists to achieve a measure of air control should the PRC attack the offshore islands again.[74]

Economic success was also a priority. As early as April 11, 1949, the well-known Nationalist financier and minister T. V. Soong (at that point in the United States) had warned that Taiwan's economy "must not be dragged down to the mainland level, and that the standard of living of the people on the island must be kept higher than any the Communists can offer."[75] But Soong was cautioned that American troops would never be sent to Taiwan and that "their own efforts will be the determining factor in Formosa and that failure to make this point clearly will inevitably result in eventual failure of our efforts to prevent Communist domination of the island."[76]

There was no fixed time limit to solving the Taiwan problem; Washington's commitment to Taipei was open-ended. As one planning document from December 1956 forecast, "Until the communists stop being communists and come back into the family of nations, a continuation of substantial United States aid to Free China will be necessary."[77] Aside from the transfer of high-tech naval equipment, the U.S. decision to train better pilots for the ROC air force and to provide it Sidewinder air-to-air missiles would turn out to be, probably, the single most important addition to the Nationalist military's capabilities. In any case, the Nationalists were well prepared when the second Taiwan Strait crisis broke out in 1958.

Notes

1. Acheson to Souers, U.S. Policy Regarding Trade with China, November 4, 1949.

2. "Security of Hong Kong—Effects of Possible Future Developments in Formosa," Report by the Chiefs of Staff, November 18, 1949, Top Secret, CAB 21/3272, TNA.

3. "Chinese Communist Naval Vessels," *ONI Review* (July 1951), pp. 271–74.

4. Central Intelligence Agency, Probable Developments in Taiwan, June 16, 1949.

5. John Foster Dulles, telegram to President Eisenhower, May 10, 1955, Top Secret, DH 5, JFD May 55, DDEPL.

6. ComTaiwanDefCom (US)/MAAG Taiwan, telegram to CinCPac, September 2, 1958, p. 1, Top Secret, DDE AWF, Int. Series 11, Formosa (1), DDEPL.

7. Memorandum of Conference with the President, September 30, 1958, Secret, DDE Diary 36, Staff Notes Sept. 58, DDEPL.

8. Memorandum for the President, January 18, 1951, Top Secret, PHST, President's Secretary's File, box 188, P.S.F. Subject File, HSTPL.

9. Summary of Telegrams, December 31, 1948, Top Secret, PHST, SMOF–Naval Aide, box 23, State Department Briefs File, HSTPL.

10. General Gruenther to Eisenhower, February 8, 1955, DDE Ad. Series 16, Gruenther, General Alfred 1955 (4), DDEPL. The slang, arising from the then-popular comic strip *Joe Palooka,* has the sense of "clumsy, oafish brawler."

11. Abbott Washburn, Deputy Director USIA, to President Eisenhower, September 9, 1958, p.

3, Secret, DDE Ad. Series 29, Quemoy Matsu, Washburn, Abbott, DDEPL.

12. Dulles, Draft "Formosa" Paper to Eisenhower, pp. 9–10.

13. Eisenhower to Dulles, February 21, 1955; responding to Dulles to Eisenhower, February 21, 1955.

14. Memorandum for the Secretary of State, April 5, 1955, p. 10.

15. Dulles, Draft "Formosa" Paper to Eisenhower, pp. 17–18.

16. Telephone Conversation between Eisenhower and Radford, February 1, 1955, DDE Diary Series 9, Phone Calls, Jan.–July 1955 (3), DDEPL.

17. Dulles, Draft "Formosa" Paper to Eisenhower, p. 11.

18. Memorandum for the Secretary of State, April 5, 1955, p. 6 [emphasis original].

19. President Eisenhower to General Gruenther, February 1, 1955, Top Secret, DDE Ad. Series 16, Gruenther, General Alfred 1955 (4), DDEPL [emphasis original].

20. Ibid.

21. Meeting of Secretary with Congressional Leaders, January 20, 1955, p. 10.

22. Telephone conversation between Eisenhower and Dulles, February 16, 1955, DDE Diary Series 9, Phone Calls, Jan.–July 1955 (3), DDEPL.

23. Dulles, Draft "Formosa" Paper to Eisenhower, p. 11.

24. Telephone conversation between Eisenhower and Dulles, February 16, 1955.

25. Memorandum of Conference with President, February 19, 1955, Top Secret, DDE Papers 4, ACW Diary, Feb. 1955 (2), DDEPL.

26. Eisenhower to Dulles, February 21, 1955; responding to Dulles to Eisenhower, February 21, 1955.

27. Breakfast Discussion with Secretary of State John Foster Dulles, July 22, 1955.

28. President Eisenhower to Prime Minister Churchill, January 25, 1955, Top Secret, PREM 11/867, TNA.

29. Message for Transmittal to the Prime Minister, February 19, 1955.

30. President Eisenhower to L. W. Douglas, March 9, 1955, Personal and Confidential, DDE Diary Series 10, DDE Diary March 1955 (2), DDEPL.

31. President Eisenhower to L. W. Douglas, March 29, 1955, pp. 1, 3, Personal and Confidential, DDE Diary Series 10, DDE Diary March 1955 (1), DDEPL [emphasis original].

32. President Eisenhower to L. W. Douglas, April 12, 1955, DDE Diary Series 10, DDE Diary April 1955 (1), DDEPL [emphasis original].

33. Immediate Actions re the Formosa Situation, March 19, 1955, pp. 2–3, Top Secret, Dulles, J. F., WHM, box 2, WHM 1955, Formosa Straits (2), DDEPL.

34. Memorandum for the Secretary of State, April 5, 1955, p. 9.

35. Dulles, telegrams to Eisenhower, February 25, 1955.

36. Dulles, Draft "Formosa" Paper to Eisenhower, p. 7.

37. Eisenhower to Churchill, March 29, 1955, pp. 4–5.

38. Warren I. Cohen, ed., *New Frontiers in American–East Asian Relations* (New York: Columbia Univ. Press, 1983), pp. 150–52.

39. Memorandum of Conference with the President April 1, 1955, April 4, 1955, Top Secret, DDE Papers 5, ACW Diary, April 1955 (6), DDEPL.

40. Memorandum for the Secretary of State, April 5, 1955, pp. 7–8.

41. "Memorandum for the Record, by the Ambassador in the Republic of China (Rankin)," April 29, 1955, in *China,* ed. Harriet D.

Schwar, Foreign Relations of the United States [hereafter FRUS], 1955–1957, vol. 2 (Washington, DC: U.S. Government Printing Office [GPO], 1986), pp. 529–31.

42. Untitled draft, April 18, 1955, pp. 4–5, Top Secret, Dulles, J. F., WHM, box 2, Offshore April–May 1955 (1), DDEPL.

43. Asst. Secretary of State Robertson, telegram to Secretary of State Dulles, April 25, 1955, sec. 1, p. 3, sec. 2, p. 2, Top Secret, DDE AWF, Int. Series 10, Formosa (China) 1952–57 (4), DDEPL.

44. Anderson, *Reminiscences,* p. 280.

45. Robertson to Dulles, April 25, 1955, p. 1.

46. Memorandum of Conversation, President's Luncheon for Sir Anthony Eden at the White House, January 31, 1956, p. 3, Secret, DDE Diary 12, Jan. 56 Goodpaster, DDEPL.

47. "Memorandum for the Record, by the Ambassador in the Republic of China (Rankin)," April 29, 1955.

48. Memorandum of Conversation, January 31, 1956, p. 3.

49. Eisenhower, letter to Dulles, April 26, 1955, pp. 1–2, Top Secret–Eyes Only, JFD Papers, WHM 8, Corres. Pres., Personal 1954–58, DDEPL [emphasis in original].

50. Memorandum for the Record, May 3, 1955, Top Secret, DDE Papers 5, ACW Diary, May 1955 (7), DDEPL.

51. Arthur Radford, Chairman Joint Chiefs of Staff, telegram to CinCPac Pearl Harbor, May 5, 1955, Top Secret, DDE AWF, Int. Series 10, Formosa Area, U.S. Mil. Ops. (1), DDEPL.

52. CinCPac/CinCPacFlt, Intelligence Estimate, n.d. [probably mid-1955], p. 2, Secret, DDE AWF, Int. Series 9, Formosa Visit 1955 (3), DDEPL.

53. Ibid., p. 17.

54. President Eisenhower to A. C. Wedemeyer [Gen.], February 28, 1955, Confidential, DDE Diary Series 9, DDE Diary, February 1955 (1), DDEPL.

55. Dulles, telegrams to Eisenhower, February 25, 1955.

56. Memorandum for the Secretary of State, April 5, 1955, pp. 2, 4 [emphasis original].

57. M. Shoosmith to Headquarters, United Nations Command, Office of the Deputy Chief of Staff, March 30, 1953, Secret, FO 371/105323, TNA.

58. President Eisenhower to Prime Minister Churchill, February 10, 1955, Top Secret–Eyes Only, PREM 11/879, TNA [emphasis original]; also in DDE Diary Series 9, DDE Diary, February 1955 (2), DDEPL.

59. Eisenhower to Churchill, February 19, 1955.

60. JFDP, reel 210/211, 93578, June 29, 1955, Princeton Univ.

61. Memorandum for the President, December 18, 1954.

62. Arleigh A. Burke, *Recollections of Admiral Arleigh A. Burke,* Oral History 64, NIOHP, pp. 43–44.

63. Robert B. Carney [Adm.], CNO, memorandum to Joint Chiefs of Staff, "Security of the Offshore Islands Presently Held by the Nationalist Government of the Republic of China," August 20, 1953, Top Secret, Strategic Plans Division, box 289, NHHC.

64. Memorandum for the President, December 18, 1954.

65. Robert B. Carney [Adm.], CNO, memorandum to Joint Chiefs of Staff, "Review of Chinese Nationalist Forces," December 16, 1953, Top Secret, Strategic Plans Division, box 289, NHHC.

66. NSC, Statement of Policy, "U.S. Policy toward Formosa and the Government of the Republic of China," January 15, 1955, Top Secret, Top Secret, WH Office, OSANSA, NSC Subseries 14, NSC 5503–Pol. toward Formosa and ROC, DDEPL.

67. NSC, U.S. Policy toward Taiwan and the Government of the Republic of China, October 4, 1957, p. 1, Top Secret, WH Office, OSANSA,

Records 1952–61, NSC Series, Policy Papers Subseries, box 22, DDEPL.

68. NSC, Statement of Policy, January 15, 1955.

69. Philip A. Beshany, *The Reminiscences of Vice Admiral Philip A. Beshany,* Oral History 45, NIOHP, pp. 887–88.

70. "Summary of NDAP Support," March 24, 1955, Secret, Strategic Plans Division, box 326, NHHC.

71. Report on Foreign Economic Policy Discussions between United States Officials in the Far East and Clarence B. Randall and Associates, December 18, 1956, p. 7, Secret, DDE U.S. Council on For. Econ. Policy, Randall Series, Trips Subseries, box 2, Far East Trip [Dec. 1956], Final Report, DDEPL.

72. Gabe T. Wang, *China and the Taiwan Issue: Impending War at Taiwan Strait* (Lanham, MD: Univ. Press of America, 2006), p. 160.

73. Eisenhower to Churchill, March 29, 1955, p. 5.

74. Anderson, *Reminiscences,* pp. 310–11.

75. Summary of Telegrams, April 11, 1949, Top Secret, PHST, SMOF–Naval Aide, box 23, State Department Briefs File, HSTPL.

76. Summary of Telegrams, April 18, 1949, Top Secret, PHST, SMOF–Naval Aide, box 23, State Department Briefs File, HSTPL.

77. Report on Foreign Economic Policy Discussions, December 18, 1956, pp. 7–8.

CHAPTER FIVE

The Second Taiwan Strait Crisis, 1958

An oppressed army fighting with desperate courage is sure to win.
(哀兵必胜.)

Tensions between the PRC and Taiwan remained high after the first Taiwan Strait crisis, in 1954–55. The Nationalist blockade of the PRC continued, albeit at a reduced level after the Dachens were evacuated. The blockade halted a lower percentage of the PRC's international shipping, since its operation was more strictly limited to southeast China. However, the combined effects of the blockade and of the U.S.-sponsored strategic embargo, which was to last until 1971, were significant.

To make up for the loss of international seaborne commerce, the PRC was at first forced to turn to the Soviet Union, conducting an ever larger share of trade via the Trans-Siberian Railway. By the late 1950s, however, it found itself overreliant on the USSR. China's debts to the USSR were well over a billion U.S. dollars—by one estimate 1.5 billion rubles (almost two billion 1962 dollars).[1] At about the same time, there was increasing opposition from the British, Japanese, Germans, Canadians, and French to the China sanctions. In early May 1957, the British argued that the "China differential should be completely abolished."[2] Dulles wrote to Foreign Secretary Selwyn Lloyd, offering to try to meet Britain halfway but insisting that "in our opinion, this differential has a real significance in retarding the buildup of Communist China's vast military potential."[3] A similar message from Eisenhower to Prime Minister Harold Macmillan warned "that many of the items which you would take off the China list will in fact appreciably help the Chinese Communists to build up the military potential which threatens us in this area and which we have the primary responsibility to resist."[4]

In the midst of all this, the Nationalists were retightening the blockade. The second Taiwan Strait crisis, during 1958, arose from the PRC's goal of halting the blockade once and for all and thereby freeing itself to diversify its overseas trade away from the USSR. These international economic initiatives were matched by domestic ones, intended to catch up with the West. Some of these policies were unsound, such as Mao Zedong's 1958 "Great Leap Forward," which eventually produced a nationwide famine that killed millions. To these ends Beijing renewed attacks on Jinmen—the Nationalists' main blockade base—and so put extreme pressure on Taipei.

The Gradual Decline of the Nationalist Blockade

During the early 1950s, the U.S. government supported the Nationalist blockade, in particular when its enforcement strengthened its own strategic embargo of the PRC. Over a year and a half, from early 1954 to mid-1955, there were thirty-five reported incidents against British shipping; the number of serious Nationalist attacks dropped to a total of fourteen, however—nine from the sea, two from the land, and three from the air—with no reported deaths or other casualties.[5] The British government, which announced in 1957 that it was planning to increase trade with China, continued to oppose the blockade with its own Formosa Straits Patrol, though that force operated more sporadically than its American counterpart during the mid-1950s. Usually one Royal Navy ship was on station at a time, and each patrol only lasted two or three days. During a five-month period in late 1954, for example, only seven ships patrolled the Taiwan Strait, each for two or three days, which meant that Royal Navy ships were present on only twenty-four out of about 150 days, or about 16 percent of this period.[6]

Unlike the U.S. Navy, whose Taiwan Patrol Force had complex duties that included patrolling, training, and morale building, the Royal Navy sought simply to protect British shipping from interference by the Nationalists and their guerrilla allies. One sure sign of the British patrol's impact, despite its light "footprint," appeared on September 8, 1955, when the Nationalist government's department of defense ordered that "attacks on shipping off the coast of China must in future be confined to Communist vessels, and that no (repeat no) neutral ships are to be molested unless this is 'unavoidable in the inherent right of self-defence.'" This sole exception, "self-defence," would appear to have envisioned a hypothetical situation where a neutral vessel "happened to be in the way of a *bona fide* attack on a Communist vessel."[7]

That spring Washington tried its best to de-escalate matters. Dulles told Eisenhower after a March 1956 visit to Taiwan that he had not found "any feeling that an all-out Chicom assault was likely in the early future. . . . From my talks with the US Country Team [the ambassador's key subordinates] I think there is a somewhat excessive tendency on the part of the Chinats to aggravate the situation by minor plane and artillery initiatives, and I think we should try to bring this under closer control."[8] Although British-flagged vessels continued to be stopped and searched from time to time, during July 1956 the British consulate in Taiwan reported that so far that year no British ship had "sustained damage or casualties as a result of Nationalist air and naval action or by shore batteries from coastal islands."[9] Almost a year after that, in May 1957, it was further reported that since December 1955 "no British ships have been damaged and there have been no casualties."[10]

Nevertheless, and despite numerous U.S.-U.K. talks over several years, there was no agreement on how to adjust the trade-restriction regime against China. On May 24, 1957, Eisenhower sent a final plea to Macmillan not to change the Chinese trade differential:

> As an individual I agree with you that there is very little of profit in the matter either for your country or for any other. Commercially, it affects this nation not at all, for the simple reason that we have a total embargo on Chinese trade. However, many of our people think that the free nations could make a terrific psychological blunder in this matter and possibly even lose all the areas of the Southeast [Asia] that have strong Chinese minorities.

> We understand your predicament and even though we may be compelled, in the final result, to differ sharply in our official positions, I think that each of our Governments should strive to prevent the possible popular conclusion in its own country that we are committed to going "separate ways."[11]

Macmillan nonetheless announced that Britain would diversify its trade with China, shipping a wider range of goods. In response, the Nationalist government intensified its blockade. On June 7, 1957, the Nationalist minister for foreign affairs pledged that his "Government would stand firm on its mainland port closure order whether or not Britain used warships to escort merchant ships sailing into Communist ports."[12] On June 15 the U.S. government cautioned Taiwan: as a White House staff member minuted, "We are advising the GRC [Government of the Republic of China] that while we recognize the importance to them of preventing the shipment of strategic materials to Communist China, their own interest calls for caution in intercepting foreign commercial shipping, particularly British, in the Taiwan Strait. We shall point out that a serious incident involving a free world ship could seriously hurt the Nationalists."[13]

Beginning in early 1958, both the total size of the Sino-British trade and the number of items traded began to increase dramatically. The U.S. embassy in Taipei had already warned, during summer 1957, that "the Chinese Communists may wish to neutralize it [Jinmen] in order to facilitate a greater use of the harbor following the British action on trade controls."[14] The PRC did indeed attempt to "neutralize" Jinmen, and the result was the second Taiwan Strait crisis.

The Chinese Shell Jinmen

As the U.S. embassy had predicted, one of the PRC's top priorities in 1958 was to increase its trade with Britain. The key to making this new policy effective was to undermine the effectiveness of the Nationalist base on Jinmen. Since the mid-1950s an estimated 750,000 PLA troops had been permanently stationed along the mainland coast opposite the offshore islands. This deployment was a constant drain and "definitely slowed down military probing that Communists might otherwise have been inclined to do."[15] That August, Prince Norodom Sihanouk, the president of the Council of Ministers of Cambodia, visited China to meet with Mao Zedong and Zhou Enlai. As he explained in mid-September to Walter S. Robertson, assistant secretary to the U.S. mission to the UN, the PRC leaders were "concerned by the fact that the offshore islands are being used to mount Commando attacks on the mainland and to impose a blockade."[16] Accordingly, and although the British decision to liberalize trade with China was an important contributing factor, the PRC's "first objective" during the second Taiwan Strait crisis "was to deter the Nationalists from using the offshore islands for harassment of the mainland, or as a base for a future invasion of the mainland."[17]

On August 23, 1958, Communist forces began shelling Jinmen, firing an estimated forty thousand shells during the first attack. Dulles told Eisenhower that even an early estimate of twenty-five thousand rounds seemed "exaggerated" but that even "if it was only half that it would still be serious."[18] The Nationalists immediately requested full U.S. military support. But Eisenhower rejected the notion that the United States had no choice but to defend the offshore islands: "The President said there is no military reason for the Chinese Nationalists to hold the offshore islands, just as there is no military advantage that the Chinese Communists would gain from an attack on Taiwan. However, we have to take our decision not on the basis of our military evaluation, but on an evaluation of the moral[e] factor."[19] Dulles agreed with this view, pointing out the contrast with the criticality of holding Jinmen and Mazu; if those two islands "were lost, the Chinese Nationalists do not consider that they could hold Formosa. Morale would crumble and Chiang's control would be lost. . . . The loss of Formosa would in his opinion be a mortal blow to our position in the Far East."[20] There was also great fear that the Soviets would become involved: "The President said we should not be drawn into spreading out the area of conflict, and thereby probably bringing the USSR in to render support to its principal ally, thus leading to general war. We must try to define fixed limits to the action."[21]

On August 27, 1958, Chiang Kai-shek sent Eisenhower a long letter describing the PRC attack on Jinmen: the shelling, strafing by MiG fighters, and the "sinking and damaging of two vessels" that had been evacuating wounded. He specifically asked for three actions: an American declaration invoking the Formosa Resolution; assignment of the U.S. Navy to convoy vessels from Taiwan to Jinmen and Mazu; and delegation to the local MAAG of "appropriate authority to make decisions when the United States and Chinese Governments hold consultations on military operation in pursuance of the exchange of notes following the signing of the mutual defense treaty." He also reminded the president that he had upheld his own part of the 1954 secret exchange of notes: "I am sure that Your Excellency is aware of the fact that we have never once made any provocative move against the Communists in the Taiwan straits during the past three years because of the Sino-American mutual defense treaty relationship."[22]

Nevertheless, Eisenhower was determined not to escalate the incident. Of Chiang's claims that the Nationalists were too weak, it was "suggested" among the president's advisers "that Chiang may be using this line to gain leverage on us."[23] As for invoking the Formosa Resolution, on August 29, 1958, President Eisenhower told Under Secretary of State (and recent governor of Massachusetts) Christian Herter that "he did not wish to put ourselves on the line with a full commitment. The Orientals can be very devious; they would then call the tune."[24] Vice Adm. Roland N. Smoot, the commander of the U.S. Taiwan Defense Command from 1958 to 1962, had no choice therefore but to tell Chiang that in accordance with the U.S.-ROC defense treaty "we would not be directly involved in this affair."[25]

The PRC attack on Jinmen did eventually draw in the Seventh Fleet, but only in a supporting role. A USIA foreign-public-opinion poll determined that "Nationalist China would greatly increase its world stature if it were able to handle the [PRC] threat itself."[26] Gen. Maxwell Taylor and Admiral Burke reported to Eisenhower that Chiang had already allocated "one-third of the effective Chinese Nationalist forces" to the offshore islands and that the Nationalists, "given the will[,] . . . could hold out against interdiction." Artillery alone could not push the Nationalist troops off the islands: "No amount of artillery fire nor of bombing would by themselves occupy the Islands for the Communists; therefore a major element in the defensibility of the Islands is the ability and especially the willingness of the 80,000-odd Chinese Nationalist forces on them to continue their defense." The U.S. Navy could support the Nationalists, but its commander's "authority did not include operations inside the three mile limit off the China coast."[27]

From the Nationalist point of view, a three-mile limit left a sizable area of China's littorals unprotected—warships could shell targets ashore from beyond it. The United States and China disagreed on how far out China's sovereignty extended: the Americans insisted on three miles, but Beijing declared a twelve-mile limit. CinCPac, then Adm. Harry D. Felt, wanted to make "occasional, purposeful intrusions so as to indicate our nonacceptance of this limit." Secretary Dulles and General Twining, however, disagreed and "pointed out that our position in the Warsaw conference makes our attitude clear and there is no need to demonstrate our attitude by overt acts at this time."[28] This was the policy being carried out at the time with regard to the Soviet Union; the only exception was a "certain discretion given to Admiral Smoot to meet unpredictable emergencies." General Twining assured the president that "he did not conceive an emergency that could occur that would not, in fact, permit prior consultation with Washington, but he still thought some discretion was appropriate."[29]

Indirect American Support for Taiwan

Though U.S. naval forces were ordered not to operate within three miles of the Chinese coast, the American military did assist Taiwan in other ways—for example, with naval "demonstrations." But even these nominally noncombatant operations could be costly in terms of aircraft and pilots. In September 1958, for example, carrier air "group commanders were a little too enthusiastic," and the "Navy paid a price for the show of force put on by the combat air patrol over the Taiwan Strait, losing four planes and three pilots in accidents."[30] In the event, "We could and did take over military defense of Taiwan itself, thus releasing his [Chiang's] military forces to defend and resupply the offshore islands."[31]

The secretary of state's September 4, 1958, assessment was that the PRC attack on the offshore islands was just the first step of a larger campaign that Beijing had been preparing "for over the past 3 years . . . with Soviet backing":

> The program has been begun by intense pressure on the weakest and most vulnerable of such positions, namely, the Chinat-held offshore islands of Jinmen and Mazu. It seems that the operation is designed to produce a cumulating rollback effect, first on the offshore islands, and then on Taiwan, the "liberation" of which is the announced purpose of the present phase. The "liberation," if it occurred, would have serious repercussions on the Philippines, Japan, and other friendly countries of the Far East and Southeast Asia.

The offensive against the offshore islands was to be "primarily military," but the follow-up "might be primarily subversive," although "armed Chicom attack against Taiwan is not to be excluded."[32]

To make sure the PRC could not successfully attack Taiwan militarily, by mid-September 1958 the U.S. Navy had positioned five carriers and their escorts near Taiwan, and two more were on their way. Additionally, a clear message was sent to the PRC with the revelation on October 1 that a number of eight-inch howitzers capable of firing nuclear shells had been delivered to Jinmen Island.[33] As a top priority the Americans were "concentrating on getting the F-86s on the islands into operational status quickly."[34] They were also providing Taiwan with high-tech equipment by means of cargo planes. On September 8, 1958, the first of twelve F-104 Starfighters (supersonic interceptors) left Hamilton Air Force Base in Novato, California, on a C-124 cargo plane. According to Admiral Felt, it was the "first time it had ever been done, I guess. They took the little old stub wings off of them and flew them out to Taiwan, unloaded them, stuck the wings on, and there we had an F-104 squadron!"[35]

The F-104 held world records for both altitude (over ninety thousand feet) and speed (1,220 knots), and it was expected that "this deployment will be a real test of its combat capability."[36] (Eventually Taiwan acquired 247 of them, mainly from the United States but also secondhand from Belgium, Canada, Denmark, Germany, and Japan.) One of the F-104's primary missions was to patrol the Taiwan Strait, engaging PRC fighters, primarily MiG-15s. On September 30, 1958, a Navy situation report estimated that MiG losses since August 14 totaled twenty-five destroyed, five probably destroyed, and eleven either damaged or "possibly destroyed." Meanwhile, the Nationalists had only lost one F-86, and that was "due to mechanical difficulty while returning to home base."[37]

The U.S. Navy trained Nationalist crews on board their ships and Nationalist pilots at the Naval Ordnance Test Station at China Lake, California (in the use of the Sidewinder missile in combat). The Air Force too assigned experts to train Nationalist pilots. According to Admiral Felt,

> We had an Air Force section of the MAAG down there, the Military Assistance Group, which trained our Chinese friends and they were well trained, every bit as good tactically as the U.S. Air Force or Naval Air fighters. They'd go out on these patrols, out over the straits, and just loiter at their best fighting altitude, more or less presenting themselves as bait. The Chinese would come out at higher altitudes and finally couldn't resist the temptation to come down, and when they came down they got took. Also it was the first combat introduction of the Sidewinder, which had been given to the Chinese. I can't remember the numbers, but I think it was something like 21 of the Communist planes shot down and success for the Sidewinders, not 100 percent but a very fine performance.

In one air battle, on September 24, 1958, Nationalist F-86s shot down an impressive ten MiGs, with two other "probable" hits, without sustaining a single loss.[38] These were the first-ever "kills" by Sidewinders;[39] in fact, the engagement marked "the first use of any American air-to-air guided missile in actual armed engagement," and a total of 360 Sidewinders were allocated for the Chinese Nationalist air force.[40] As a result of superior U.S. equipment and training, the Nationalist pilots established air control so effectively that the "Red Chinese weren't much interested in challenging in the air."[41] Very soon, American assistance also gave the Nationalists equally potent sea control.

Keeping the Sea-Lanes Open

Getting supplies to Jinmen was the top military priority, and the U.S. Navy's most important contribution to it was helping protect the sea-lanes. The Nationalists were expected to take the supplies all the way to the island, but through late August they did a poor job. For example, if fired on by the Communists many Nationalist resupply ships turned around and headed back to Taiwan rather than "waiting over [the] horizon awhile and then trying again." The Navy was limited by orders from Washington to convincing the Nationalists "that [the] U.S. is willing to help them but cannot be expected to assume further responsibility for getting supplies all the way ashore on the islands unless GRC has first demonstrated that they themselves have the real determination to see the action through to the finish, despite the hazards involved, including artillery fire at the beaches."[42] In September the head of the MAAG complained to Chiang "that to date the Chinese Navy has made no effort to even try" to break the Communist blockade of Jinmen but that if Chiang could just "give me one Chinese admiral with a can do spirit then let *us* use a little imagination and a little guts and some tenacity" it would be possible to "stabilize this situation and thumb our noses at the Communists."[43]

The Nationalists' behavior led to speculation that they were trying to pull U.S. forces deeper into the conflict: "There is a possibility that GRC is being deliberately inept in order to draw U.S. inextricably into conflict with CHICOMs."[44] In a meeting with the president in early September the Joint Chiefs requested authority to "approve U.S. air support" for the Nationalists. Eisenhower's reply reflected his concern that the United States not be drawn unwillingly into Chiang's fight: he recalled that the chiefs had estimated such American air support would not be required unless Chinese communist air forces attacked en masse in support of land operations, and that there would be time for his decision in such cases. Accordingly, he prescribed that U.S. air attack against mainland targets could be ordered only upon his approval.[45]

Also in early September, the Taiwan Patrol Force was ordered to provide landing ships and craft, escorts, and support forces to protect the Nationalist resupply vessels. They were warned to stay in international waters—that is, "beyond the three-mile limit." Additional equipment was being rushed to the scene, including eight medium landing craft (LCMs) and twenty-eight vehicle and personnel landing craft (LCVPs), a squadron of Sidewinder-equipped F-100 Super Sabre fighters, and twelve nuclear-capable eight-inch howitzers plus ammunition.[46] On September 6, the first U.S.-escorted Nationalist convoy, code-named LIGHTNING, reached Jinmen with crucial supplies. By September 19 nine convoys had been conducted, each of the final four landing, on average, 151 tons. By mid-September Nationalist supply ships had, with U.S. Navy assistance, overcome what was being called a PRC "artillery blockade" of the island. By the end of September 1958, Jinmen had at least thirty-one days' worth of food and thirty-two of ammunition on hand. Its endurance in other categories—such as spare parts—was twice that; the average had increased from fifty days to fifty-five during the past two weeks alone.[47] The Joint Chiefs reported to the president that the supply crisis on the offshore islands was now "broken."[48]

To safeguard the supply ships' arrival and withdrawal from the contested area, U.S. aircraft flew antisubmarine patrols and surface reconnaissance out to twenty-five miles from Jinmen, staying at least twenty miles off the Chinese coastline. U.S. warships were ordered to remain at least three miles from shore and were particularly warned not to "shoot at the mainland."[49] However, a special "rule of engagement" authorized "US Commanders . . . to engage hostile surface vessels in territorial or international waters if they are attacking the [Republic of China Navy] forces."[50] Meanwhile, American naval personnel trained the Nationalists intensely to carry out successful convoy operations. A map from September 15, 1958 (map 2), shows how U.S. Navy ships—well outside China's sovereign waters, as indicated on the map by dotted boxes—were stationed to protect the Nationalist ships resupplying Jinmen.

Notwithstanding, Washington's support for the Nationalists with regard to the offshore islands remained conditional. About four months before the crisis broke out, in May 1958, the NSC, in a lengthy report on Taiwan, had reiterated that "loss of Taiwan and the Penghus would out-flank the United States base on Okinawa and would seriously

MAP 2
U.S. Navy's Jinmen Convoy Operations

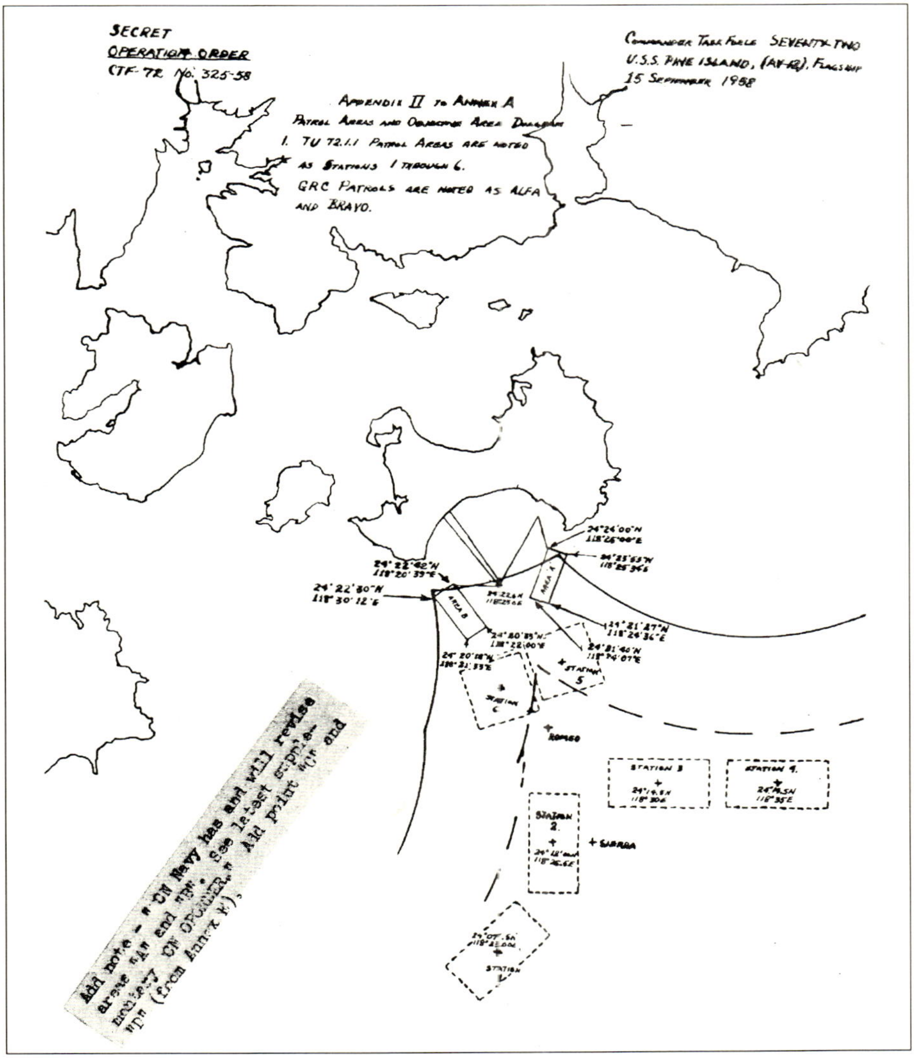

breach the Free World's defense line in the West Pacific, which stretches from South Korea and Japan to the Philippines." This same study, however, hypothesized that when attacked by the PRC, Chiang would evacuate "Quemoy and Matsu" but "only after strongest U.S. representations" and that during the evacuation there might be "some casualties and ship damage due to mines."[51]

Washington continued to refuse Taipei a blank check. For example, in August Chiang Kai-shek wanted to use Taiwanese planes to bomb the mainland, but Washington expressed concern that this might escalate the conflict and draw U.S. forces into it. According to Admiral Smoot, a Navy study had proved that for every gun the Nationalists destroyed they might lose a squadron of planes: "This, of course was too big a price to pay, and they [Smoot's staff] were convinced of the proposal's infeasibility."[52] Chiang was not pleased by America's decision not to conduct "retaliatory air attacks against Communist air fields, et cetera," calling its attitude "inhuman" and "unfair" to his soldiers on the offshore islands and generally "destructive of public morale." Ambassador Everett F. Drumright in Taipei reported that Chiang's "reaction was the most violent I have seen him exhibit and at one point he called our policy 'not that of an allied nation.'"[53]

In early September 1958, the New Zealand embassy in Washington, acting on behalf of the U.S. government, appealed to the Soviet ambassador, Mikhail Menshikov, "to restrain" Beijing.[54] Late that month the PRC sent a message through Indian intermediaries that if the Nationalists withdrew from the islands they would not be attacked in the process and that leaders in Beijing were "not concerned to press immediately their claims to Taiwan."[55] The Chinese were, however, clearly concerned about further U.S. intervention; they warned their artillery not to aim at American ships. However, Mao Zedong refused to accept American demands that a cease-fire precede Sino-U.S. talks to resolve the crisis. For this reason, a negotiated settlement appeared unlikely.

By the end of October 1958, the convoying situation had been largely resolved; the Communists now allowed the Nationalists to send in convoys on even-numbered days. As a result, the Navy's tasking changed: "U.S. Commanders are instructed to convoy only in case of military necessity, which would be limited to situations where the Communists attack supply convoys on even numbered days." So long as the Nationalist supply ships were not attacked, U.S. ships did not need to escort them at all. Admiral Smoot had agreed to the even-numbered-day plan but Admiral Felt had disagreed, and "in light of political considerations, Secretary Dulles [had sided] with Admiral Felt." But the president had decided that "pending further instructions, [the Navy] would not engage in convoying operations unless the Chinese Communists attempted by sea and/or air to interfere with the re-supply on even days in what we regard as international waters."[56] However, not wanting to force the Nationalists to "jump through a hoop" at the PRC's bidding, the United States would support them if "they might feel a political reason to attempt some resupply on the days when the Chicoms had announced they would fire."[57]

Dulles Convinces Chiang to Withdraw

One way to de-escalate the crisis was to get the Nationalist troops off the offshore islands. The American and British governments were in constant consultation. On September 9, 1958, the British chargé in Beijing, Sir Duncan Wilson, warned that to obtain "great power status" the PRC was willing to foment foreign crises, that the "Chinese Government seems thoroughly arrogant and overconfident in their diplomacy." The Chinese leaders' staying power was almost limitless, since they were "ready, if necessary, to sacrifice enormous numbers of men in limited military operations" and would deliberately "extend" a limited war rather than look defeated. Wilson argued, with his "Dutch, Norwegian, Danish and Pakistan colleagues" in agreement, that to resolve the ongoing crisis the PRC must "have at least a diplomatic victory to show their people" to cite as an excuse to "back down."[58]

Washington was more concerned than ever that fighting in the Taiwan Strait might escalate into all-out war. Further, as Dulles told Eisenhower in August 1958, Jinmen and Mazu were now so integrated with Taiwan and the Penghus that he "doubted whether there could be an amputation without fatal consequences to Formosa itself."[59] On October 13 General Twining reported to Eisenhower that the Joint Chiefs and Dulles had just concluded that once the PRC artillery fire "quiets" it would be necessary to "withdraw at least two-thirds of the Chinese Nationalist Troops." Eisenhower agreed but cautioned that the next step was to convince Chiang Kai-shek of it: "We must accomplish our end through persuasion rather than pressure, since otherwise we will be charged with reversing the stand we have taken." In response, Twining proposed that Chiang be offered American assistance to "modernize his Army; we could partially modernize his Air Force, although this would be very expensive; we could give him shipping and amphibious lift to enlarge his capabilities for flexible action; perhaps we could even give him additional economic aid."[60]

Dulles summarized for file a conversation he had with Ambassador Lodge on September 19, 1958. Of his goals with regard to Chiang Kai-shek,

> I said that we had tried repeatedly to get the [Taiwanese] government to withdraw or at least greatly reduce its forces on Quemoy and Matsu and to treat them as lightly held forward positions to be retired from if necessary but that we had constantly come up against the fact that Chiang Kai-shek was adamant in rejecting such a viewpoint as incompatible with the entire basis for his government and that if we attempted to coerce through a cut-off of military assistance and financial aid, that itself would be as destructive of a friendly position on Formosa as though it was taken over overtly by the Communists.

A second issue discussed was stopping the Nationalist attacks against the mainland. To de-escalate tensions, Beijing had been told that Washington was working to end them.

Dulles told Lodge, however, that he "was not confident that we could get the Chinese Nationalists to accept the plan to end 'provocations' from the Offshore Islands as we had proposed to the Communists."[61]

Dulles reported by telephone to Vice President Richard Nixon that repeated attempts had been made to get Chiang to give up the offshore islands, even sending Radford and Robertson to talk with him, but that Chiang had firmly said no. The dilemma was that "we can break them if we cut off aid but if you break them, you lose Formosa." The Americans could not afford to do that: "The broad challenge is are we going to keep the Western shores of the Pacific in friendly hands or not? We can fall back initially by giving up Q[uemoy] and M[atsu] which will carry with it the loss of Formosa and then we lose the Philippines, Japan will make terms with Communist China and we will have suffered an overall setback even worse than when we lost the Mainland."[62]

On September 21, 1958, the president told Foreign Secretary Lloyd of the United Kingdom that "he would be very happy indeed if the United States could make some arrangement with Chiang Kai-shek which would not lose face for him but which would get his troops off the islands of Quemoy and Matsu."[63] In a separate conversation with Dulles, the president "expressed regret that there seemed to be no way to persuade Chiang to re-direct the focus of his leadership, in a way which would enable him to re-group his military forces into more sensible positions."[64] However, Eisenhower clarified that "it was essential that the free world keep control of the island of Formosa and that if Formosa were lost, then a hole would result in the very middle of the island chain of defense. Should the Reds eventually control Formosa, that, in the President's opin-ion, would be a real Munich." If, in the worst case, "Chiang Kai-shek were to quit and Formosa went to the Reds, then the overseas Chinese would have no place to go except to the Communist camp."[65] To convince Chiang to cooperate, Dulles decided to sign a second secret agreement.

The Second Secret Agreement

Reducing forces on the offshore islands would necessarily end the blockade. During late October 1958, Dulles's task was to persuade Chiang Kai-shek to reduce the size of Nationalist forces on the islands so as to halt "commando raids and blockades."[66] Eisen-hower repeatedly wrote "strong" letters to Dulles suggesting that Chiang "be offered an amphibious lift in exchange for getting his troops out of Quemoy and Matsu."[67] Dulles flew to Taiwan to convince Chiang to withdraw from the offshore islands.[68] During private talks, however, Chiang refused to withdraw, rejecting "any proposal that seems to him to imply retreat from his position as head of the only legitimate Chinese Govern-ment."[69] Dulles nonetheless urged him "to renounce the use of force in an attempt to reunify China."[70] Doing so would mean "a substantial reduction of forces" on Jinmen and Mazu.[71]

On October 23, 1958, Dulles reported from Taipei in a telegram for the president's "eyes only" that Chiang had "accepted the principle of an appreciable reduction of forces on Quemoy to be effected whenever there was a suspension of the fighting."[72] In return for a reduction of forces on Jinmen by "not less than 15,000 men," including "1 infantry division plus additional units and/or individuals," Dulles had agreed to greater arms shipments, including "a minimum of 12 240mm howitzers" and a "minimum of 12 155mm guns" for Jinmen and four or more 240 mm howitzers plus one battalion of 155 mm guns, "when available," for Mazu. The possibility was also raised for "Lacrosse missiles to be considered at a later date," a "minimum of 1 tank battalion," and, if changes to the M-8 assault gun proved "feasible," "sufficient converted vehicles to motorize 2 battalions of infantry."[73] The Nationalists had eight months to reduce their forces on the offshore islands, and "the target date for completion of the implementation of the above agreement is June 30, 1959." That this quid pro quo was more a "face-saving" measure for Chiang than a true ROC necessity is made apparent in a comment by Oliver M. Gale, special assistant to the secretary of defense: "The materiel and equipment outlined in this agreement had already been earmarked for GRC and do not actually represent an increase in previously planned assistance."[74]

Left unstated but implicit in the agreement was that any cut in forces on the offshore islands would necessarily terminate the Nationalists' decadelong blockade of China. Certainly, this was one of the American goals that Dulles outlined to the British Foreign Office immediately before he left for Taiwan to meet with Chiang Kai-shek.[75] The blockade was no longer considered by the United States as important as before, since the PRC was already beginning to experience an economic implosion due to the Great Leap Forward. Meanwhile, it was well-known in Washington that Sino-Soviet diplomatic relations were in decline. As Gordon Chang has pointed out, Eisenhower was careful not to comment on that relationship, for fear of strengthening it. Even his memoir barely mentions the point, so as "to avoid saying anything that could hinder the emergence of the Sino-Soviet split."[76]

Dulles clearly wanted to put even more pressure on the Sino-Soviet alliance, telling newspapermen in Newport, Rhode Island, that he did not think the PRC or USSR acted in "accordance with treaty obligations" and that there "have been plenty of treaty obligations which they have evaded or broken."[77] According to a Bonn intelligence report, the Soviet premier, Nikita Khrushchev, had warned Mao about the risks of attacking the offshore islands, but Mao had dismissed the warning. Mao had been convinced that it would force a retreat, as had happened in the Dachen Islands three years before, and that "liberating" these Nationalist-held islands "was necessary for internal and external prestige reasons and that Nationalist loss of the islands would undermine their position, increase their large rate of military defections, and facilitate Communist subversive activity."[78]

Nevertheless, on October 6, 1958, the PRC halted the shelling of Jinmen, after forty-four days. Civilian casualties were 138 dead and 324 injured; dead and wounded soldiers numbered close to three thousand. In addition, an estimated seven thousand buildings on Jinmen were either damaged or destroyed.[79] Sporadic artillery fire from the PRC side would continue for the next twenty years, ending for good only in January 1979 after President Jimmy Carter and Chairman Deng Xiaoping recognized each other's governments. This barrage took place on alternate days of the week, and the shells mainly contained propaganda leaflets. An estimated one million steel shell casings were fired at Jinmen during this period, making it "the longest sustained artillery warfare in world military history."[80]

It was no "coincidence that the Nationalist naval blockade also ended during 1958, right as the time the first signs of what would soon be called the Sino-Soviet 'split' began to appear."[81] Moscow's failure to support Beijing in the offshore islands crisis undoubtedly helped Washington undermine their alliance. However, failure met all U.S. attempts to resolve PRC-Taiwan differences peacefully. On October 27 Selwyn Lloyd wrote to Dulles that the Chinese "seem to be in no hurry" to make peace with Taiwan and would "pursue their aims by whatever political means offer themselves from time to time. They do not want mediation and their ultimate goal appears to be some direct arrangement with the Nationalists." Lloyd predicted, with great accuracy from the viewpoint of a half a century later, "We are, therefore, likely to be in for a fairly long period of such tactics."[82]

Conclusions

A second secret agreement signed by Dulles and Chiang ended the second Taiwan Strait crisis. This agreement also had the subsidiary effect of ending the Nationalist blockade of the PRC. This blockade had already lasted ten years and, in combination with the ongoing U.S. strategic embargo, had exerted extreme economic pressure on the PRC. Sino-Soviet economic tensions eventually forced a major realignment in the PRC's foreign trade. But the military standoff over the offshore islands remained unresolved. During this crisis the Taiwan Patrol Force "accomplished one of the most important missions of her career by playing a major role in aiding the Chinese Nationalists."[83]

According to Dulles, the real dispute was not geography but "human wills." If the United States "seems afraid, and loses 'face' in any way, the consequences will be far-reaching, extending from Viet Nam in the south, to Japan and Korea in the north."[84] Periodic PLA attacks against Jinmen made it a "whipping boy" for Taiwan itself.[85] The PRC leaders were pursuing mainly a political, not military, strategy, playing "a 'cat and mouse' game with the offshore islands." To Dulles, therefore, the PRC announcement that the shelling of Jinmen might switch from even-numbered to odd-numbered days seemed to substantiate this assessment: "This rather fantastic statement seems to confirm our analysis of the Chinese Communist attitude as being essentially political and propaganda rather than military."[86]

Meanwhile, the U.S. government's public statements remained intentionally vague about whether it might use military force to support the Nationalist bases on Jinmen and Mazu. However, a secret December 26, 1959, operation order specified that the offshore islands "are not covered by this agreement" but acknowledged that the United States had committed itself not only to the defense of Taiwan and the Pescadores but also—most importantly—to the offshore islands of Jinmen and Mazu, "insofar as a threat to them is considered to be a threat against Taiwan and the Penghus."[87] The decision to defend any offshore island that posed a direct threat, if in PRC hands, to Taiwan arguably included the use of atomic bombs, a question that was indeed hotly argued during both the Truman and Eisenhower administrations.

Notes

1. Frank Dikötter, *Mao's Great Famine: The History of China's Most Devastating Catastrophe, 1958-1962* (New York: Walker, 2010), p. 105.

2. Staff Notes No. 107, May 4, 1957, Secret, DDE Diary 24, May 57 Diary, Staff Memos, DDEPL.

3. Draft of Suggested Message from the Secretary of State to Foreign Minister Selwyn Lloyd, May 17, 1957, Secret, JFD Papers, WHM, box 6, Meeting Pres. 1957 (5), DDEPL.

4. Draft of Suggested Message from the President to Prime Minister Macmillan, May 17, 1957, Secret, JFD Papers, WHM, box 6, Meeting Pres. 1957 (5), DDEPL.

5. "Incidents Involving British Merchant Ships off the China Coast," July 18, 1955, ADM 116/6245, TNA.

6. A. H. E. Allingham to P. Wilkinson, Far Eastern Department, Foreign Office, March 24, 1955, ADM 1/26157, TNA.

7. U.K. Embassy, Washington, DC, telegram, September 8, 1955, ADM 116/6245, TNA. The telegram informed the government in London of the Nationalist order to halt attacks on neutral shipping.

8. Secretary of State Dulles, Seoul, Korea, telegram to President Eisenhower, March 19, 1956, Secret, Dulles-Herter 6, JFD Mar. 56, DDEPL.

9. Tamsui, Formosa, telegram to Foreign Office, July 6, 1956, ADM 116/6245, TNA.

10. Tamsui, Formosa, telegram to Foreign Office, May 31, 1957, ADM 116/6245, TNA.

11. President Eisenhower to Prime Minister Macmillan, May 24, 1957, Top Secret, DDE Diary 24, May 57 Misc. (2), DDEPL.

12. Tamsui, Formosa, telegram to Foreign Office, June 9, 1957, ADM 116/6245, TNA.

13. White House, Staff Notes No. 131, June 15, 1957, Secret, DDE Diary 25, June 57 Diary Staff Memos, DDEPL.

14. JFDP, June 26, 1957, reel 217/218, 97827, Princeton Univ.

15. U.K. Consulate, Tamsui, to Foreign Office, January 27, 1958, FO 371/33522, TNA.

16. "Memorandum of Conversation," September 16, 1958, in *China*, ed. Harriet Dashiell Schwar, Foreign Relations of the United States, 1958-1960, vol. 19 (Washington, DC: GPO, 1996) [hereafter *China, 1958-1960*], pp. 201-203.

17. Joseph F. Bouchard, *Command in Crisis: Four Case Studies* (New York: Columbia Univ. Press, 1991), p. 59.

18. Telephone Calls, August 23, 1958, DDE Herter, CAH Tel. Calls, box 11, CAH Tel. Calls 7/1/58–9/30/58 (1), DDEPL.

19. Memorandum of Conference with the President, August 14, 1958, pp. 1–3, Top Secret, DDE Diary 35, Aug. 58, Staff Notes (2), DDEPL.

20. Ibid., pp. 1–2.

21. Ibid., pp. 1–3.

22. Chiang Kai-shek, telegraphed letter to President Eisenhower, August 27, 1958, DDE AWF, Int. Series 11, Formosa (China), 1958–61 (3), DDEPL.

23. Memorandum of Conference with the President, August 29, 1958, pp. 1–2, DDE AWF, Int. Series 11, Formosa (3), DDEPL.

24. Memorandum of Conference with the President, August 25, 1958, p. 2, DDE Diary 35, Aug. 58, Staff Notes (1), DDEPL.

25. Bouchard, *Command in Crisis*, pp. 76–77.

26. Staff Notes, September 3, 1958, DDE Diary 36, Toner Notes Sept. 58, DDEPL.

27. Taiwan Straits: Issues Developed in Discussion with JCS, September 2, 1958, pp. 1–6, Top Secret, DDE AWF, Int. Series 11, Formosa (2), DDEPL.

28. Memorandum of Conference with the President, October 30, 1958, pp. 1–2, Top Secret, DDE Diary 36, Staff Notes Oct. 58, DDEPL.

29. Memorandum of Conversation with the President, October 30, 1958, pp. 1–2, Top Secret, JFD Papers, WHM 7, WH Meetings J–Dec. [July–December] 1958 (3), DDEPL.

30. Bouchard, *Command in Crisis,* pp. 76–77.

31. Roland N. Smoot [Vice Adm.], "As I Recall . . . The U.S. Taiwan Defense Command," U.S. Naval Institute *Proceedings* 110/9/979 (September 1984), pp. 56–59.

32. Summary, Estimate of Factors Involved in the Taiwan Straits Situation, September 4, 1958, pp. 1–2, Top Secret, DDE 36, Staff Notes Sept. 58, DDEPL.

33. Chang-Kwoun Park, "Consequences of U.S. Naval Shows of Force, 1946–1989" (PhD diss., Univ. of Missouri–Columbia, August 1995), pp. 257–60.

34. Memorandum of Conference with the President, August 11, 1958, Secret, DDE Diary 35, Aug. 58, Staff Notes (2), DDEPL.

35. Beshany, *Reminiscences,* p. 395.

36. "F-104s to Taiwan," Staff Notes 423, September 15, 1958, DDE Diary 36, Toner Notes Sept. 58, DDEPL.

37. Memorandum for the Chief of Naval Operations, September 30, 1958, Secret, DDE AWF, Int. Series 11, Formosa (3), DDEPL.

38. Robert Keng, "Republic of China F-86's in Battle," *ARC Air: Aircraft Resource Center,* accessed March 22, 2011, www.aircraftresource center.com/Stories1/001-100/021_TaiwanF-86 _Keng/story021.htm.

39. Edward J. Marolda, "Confrontation in the Taiwan Straits," in *U.S. Navy: A Complete History,* ed. M. Hill Goodspeed (Washington, DC: Naval Historical Foundation, 2003).

40. Staff Notes No. 432, September 30, 1958, Secret, DDE Diary 36, Toner Notes Sept. 58, DDEPL.

41. Harry Donald Felt, *Reminiscences of Admiral Harry Donald Felt,* Oral History 138, NIOHP, p. 391.

42. JCS WASH DC [Joint Chiefs of Staff, Washington, DC], message to Admiral Felt, September 2, 1958, p. 1, Top Secret, DDE AWF, Int. Series 11, Formosa (1), DDEPL.

43. ComTaiwanDefCom to CinCPac, September 2, 1958, p. 2.

44. JCS WASH DC to Felt, September 2, 1958, p. 1.

45. Memorandum of Conference with the President, September 8, 1958, pp. 1–2, Top Secret, DDE Diary 36, Staff Notes Sept. 58, DDEPL.

46. JCS WASH DC, message to CinCPac, August 29, 1958, pp. 1–3, Top Secret, DDE AWF, Int. Series 11, Formosa (3), DDEPL.

47. Memorandum for the Chief of Naval Operations, September 30, 1958.

48. Memorandum of Conference with the President, September 30, 1958, Secret, DDE Diary 36, Staff Notes Sept. 58, DDEPL.

49. Smoot, "U.S. Taiwan Defense Command."

50. CTF 72 Operation Order No. 325-58, September 15, 1950, Post-1946 Operation Plans, Task Force 72, NHHC.

51. U.S. and Allied Capabilities for Limited Military Operations to 1 July 1961, May 29, 1958, pp. B2, B5.

52. Smoot, "U.S. Taiwan Defense Command."

53. Ambassador Drumright, Taipei, telegram to Secretary of State, August 31, 1958, p. 1, DDE AWF, Int. Series 11, Formosa (3), DDEPL.

54. New Zealand Embassy, note, September 3, 1958, Secret, JFD Papers, WHM 7, WH Meet J–D [July–December] 1958 (7), DDEPL.

55. Peking, telegram to Foreign Office, September 25, 1958, Secret, PREM 11/3738, TNA. This offer appears to have been offered by Zhou Enlai to Indian ambassador Parthasarathi during a meeting held Sunday, September 21, 1958.

56. Memorandum of Conference with the President, October 30, 1958, pp. 1–2.

57. Memorandum of Conversation with the President, October 30, 1958, pp. 1–2.

58. Excerpt, British Chargé, Peiping, Personal Assessment, September 9, 1958, Confidential, DDE AWF, Int. Series 11, Formosa (1), DDEPL.

59. Memorandum of Conversation with the President, August 12, 1958, Top Secret, JFD Papers, WHM 7, WH Meet J–D [July–December] 1958 (8), DDEPL.

60. Memorandum of Conference with the President, October 15, 1958, pp. 1–2, Secret, DDE 36, Staff Notes, Oct. 58, DDEPL.

61. Memorandum of Conversation with Ambassador Lodge, September 19, 1958, pp. 1–2, Top Secret, Dulles, J. F., Gen. Cor. box 1, Memos L–M (1), DDEPL.

62. Telephone Call from the Vice President, September 25, 1958, p. 1, Dulles, J. F., Tel. Conv., box 9, Aug.–Oct. 58 (3), DDEPL.

63. Memorandum of Conversation between the President, Selwyn Lloyd, Foreign Minister of the United Kingdom, and Sir Harold Caccia, British Ambassador to the United States, September 21, 1958, pp. 1–6, Top Secret, DDE Diary 36, Staff Notes Sept. 58, DDEPL.

64. Memorandum of Conversation with the President, September 23, 1958, pp. 1–2, Secret, JFD Papers, WHM 7, WH Meet J–D [July–December] 1958 (6), DDEPL.

65. Memorandum of Conversation between the President, Lloyd, and Caccia, September 21, 1958, pp. 1–6.

66. Foreign Office to U.K. Embassy, Washington, DC, October 22, 1958, Secret, PREM 11/3738, TNA.

67. Diary, October 7, 1958, DDE Papers 10, ACW Diary, Oct. 1958, DDEPL.

68. U.K. Embassy, Washington, DC, telegram to Foreign Office, October 17, 1958, Secret, PREM 11/3738, TNA.

69. President Eisenhower to Prime Minister Macmillan, September 6, 1958, Top Secret, CAB 21/3272, TNA.

70. John Foster Dulles to British Ambassador, October 25, 1958, Secret, FO 371/133543, TNA.

71. John Foster Dulles to Selwyn Lloyd, October 25, 1958, Secret, PREM 11/3738, TNA.

72. Acting Secretary for President, October 23, 1958, Eyes Only, DH 10, Dulles Oct. 1958, DDEPL.

73. The MGM-18 Lacrosse was a short-range, truck-borne battlefield ballistic missile that would enter the U.S. Army in 1959, only to be withdrawn in 1964 owing to intractable technical problems.

74. Memorandum for Maj. John S. Eisenhower, December 11, 1958, pp. 1–3, Secret, DDE Papers, White House Office, Office of the Staff Secretary, Records, 1952–61, International Series, box 3, File "China (Republic of) (1)," September 1958–April 1960, DDEPL.

75. Foreign Office to Washington, DC, October 22, 1958.

76. Gordon H. Chang, *Friends and Enemies: The United States, China, and the Soviet Union, 1948–1972* (Stanford, CA: Stanford Univ. Press, 1990), p. 331 note 24.

77. News Conference, U.S. Naval Base, Newport, RI, September 4, 1958, p. 5, JFD Papers, WHM 7, WH Meet J–D [July–December] 1958 (8).

78. Staff Notes No. 414, September 3, 1958, Secret, DDE Diary 36, Toner Notes Sept. 58, DDEPL.

79. Michael Szonyi, *Cold War Island: Quemoy on the Front Line* (New York: Cambridge Univ. Press, 2008), pp. 76–77.

80. Ryan, Finkelstein, and McDevitt, *Chinese Warfighting*, p. 167.

81. See Elleman, "The Nationalists' Blockade of the PRC," p. 142.

82. Foreign Secretary Selwyn Lloyd to Secretary of State John Foster Dulles, October 27, 1958, Secret, PREM 11/3738, TNA.

83. "Post-1946, Command File, Taiwan Patrol Force," n.d., pp. 7–8, box 784, NHHC.

84. John Foster Dulles, telegram to Harold Macmillan, September 13, 1958, Top Secret, CAB 21/3272, TNA.

85. U.K. Embassy, Washington, DC, telegram to Foreign Office, October 25, 1958, PREM 11/3738, TNA.

86. Dulles to Lloyd, October 25, 1958.

87. CTF 72 Operation Order No. 201-60, December 26, 1959, Post-1946 Operation Plans, Task Force 72, NHHC.

CHAPTER SIX

The Possibility of the Use of Atomic Weapons

Nothing is too deceitful in war. (兵不厭詐.)

To most outside observers, the forces facing each other across the Taiwan Strait in the early 1950s may have seemed unevenly matched, in favor of the United States and its allies. However, asymmetric warfare can give advantages to both sides. For example, in the Korean War enormous manpower had allowed China to send "human waves" against the United Nations troops. Therefore, on the American side the possibility of being forced to use atomic weapons, in particular against an invasion force aimed at Taiwan, was seriously discussed.

During the early 1950s, atomic bombs were often thought of as much like regular bombs. The use of the A-bomb was considered in Korea and later in Vietnam during the Dien Bien Phu crisis. With regard to Taiwan, during July 1950 Truman authorized the movement to Guam of B-29 bombers capable of carrying atomic bombs. Meanwhile, an air unit in Guam was given control of nonradioactive atomic-bomb components, the nuclear core to be provided only in an emergency. This information was leaked to the *New York Times* so as to give the PRC pause about attacking Taiwan.[1]

During times of crisis, planning sessions often included discussion on using atomic bombs. For example, on September 12, 1954, a top-secret NSC paper on the Taiwan Strait crisis warned that a war with China over the offshore islands would necessarily bring a sharp rise in tension that would "probably lead to our initiating the use of atomic weapons."[2] Clearly, that possibility was considered an important element in the defense of Taiwan.

Consideration of the Nuclear Option

It is still unclear whether the U.S. government would have actually used A-bombs in Korea or to halt a PRC invasion of Taiwan. On August 6–8, 1950, MacArthur evidently told W. Averell Harriman (Truman's special assistant and administrator of the Marshall

Plan, former ambassador to the USSR and secretary of commerce) that if the PRC attacked Taiwan, Seventh Fleet ships, fighter jets from the Philippines and Okinawa, B-29s, and other aircraft could destroy the invasion force. It would be a one-sided battle: "Should the Communists be so foolhardy as to make such an attempt, it would be the bloodiest victory in Far East history."[3]

Although use of the A-bomb may not have been specifically discussed during these meetings (Harriman had been sent to see MacArthur, then commanding UN forces in Korea and the occupation of Japan, soon after the general held a controversial meeting with Chiang Kai-shek), the fact that MacArthur mentioned the (A-bomb-capable) B-29s suggests that he had their use in mind. Even if not, the employment of atomic weapons was certainly considered later. On December 1, 1950, for example, the Chief of Staff of the U.S. Army, Gen. J. Lawton Collins, stated that if the Russians actively intervened in the Korean conflict "we would have to consider the threat of the use of the A-bomb."[4] On December 27, Secretary of State Acheson asked what would happen if Russia intervened in Korea, and was told that there would be retaliation against Port Arthur (a Russian-held port in Manchuria) and Vladivostok, "using the atom bomb."[5]

On September 12, 1954, during the first Taiwan Strait crisis, the JCS similarly recommended considering the use of nuclear weapons against China. However, when the decision to evacuate the Dachen Islands was made during January 1955, it was determined that "atomic weapons will not be used."[6] A special "Atomic Operations" annex to the Dachens operations plan specified that if the enemy attacked the evacuation forces, commanders could assume that the *"use of atomic weapons will be authorized."* A list of possible target cities was marked out when the annex was declassified, on December 29, 2011, but this statement remains: *"Other targets suspected as source of enemy attacks may be nominated by COMSEVENTHFLT for atomic strikes."*[7]

On February 21, 1955, Dulles warned Eisenhower that a Communist buildup across from Jinmen and Mazu meant these islands might soon be indefensible "in the absence of massive U.S. intervention, perhaps with atomic weapons."[8] On March 7, Dulles, recently returned from Taiwan, advised Eisenhower that to support Chiang Kai-shek's position on the offshore islands might "require the use of atomic missiles," to which the president replied that "he thoroughly agreed with this" but not with "weapons of mass destruction" (considered as larger atomic bombs or nuclear devices). Still, Eisenhower acknowledged, "with the number of planes that we had available in the Asian area, it would be quite impractical to accomplish the necessary results in the way of putting out airfields and gun emplacements without using atomic missiles."[9]

On March 10, Dulles stated at an NSC meeting that the United States might use atomic weapons against China if it attacked Taiwan: "Determination must soon be made whether in such defense atomic weapons will be tactically used. The need for such use, to make up for deficiency in conventional forces, outweighs the repercussive effect of such use upon free world nations in Europe and the Far East (especially Japan, where attempt[s] may be made to immobilize U.S. forces). U.S. and world public opinion must be prepared."[10]

The next day, Eisenhower summarized the situation during a meeting at which both Dulles brothers, Radford, Twining, Carney, and Gen. Andrew J. Goodpaster (the president's defense liaison officer) were present. He declared, according to the aide-mémoire, that

> the U.S. should do every practical thing that could be done to help the Chinats to defend themselves; that if it was necessary later for the U.S. to intervene, it should do so with conventional weapons; that the U.S. should improve the air defense of the Formosa air fields, but should avoid greatly augmenting U.S. troops on Formosa; that we should give the best possible advice and training to the Chinats about how to take care of themselves; that he recognized that if we had to intervene with conventional weapons, such intervention might not be decisive; that the time might come when the U.S. might have to intervene with atomic weapons, but that should come only at the end, and we would have to advise our allies first.[11]

On March 15, Admiral Stump told the president that at the current levels the Communist air force could be held off by "U.S. *conventional* operations" but that if the PRC moved "air forces in strength into the area, the U.S. would have to be prepared to employ atomic weapons."[12] Stump later specified that if airfields deep in Fujian Province were used to attack the offshore islands or Formosa, the "danger [they] would pose to the U.S. fleet units would require that [they] be destroyed" with "special weapons"—that is, atomic weapons.[13]

At this point, atomic weapons were considered as being just one option in a large arsenal and as fundamentally like the others. Eisenhower publicly confirmed that "A-bombs can be used . . . as you would use a bullet." About ten days later, on March 25, the CNO, Admiral Carney, stated that the president was planning to destroy Red China's "military potential," which certainly implied possible use of the atomic bomb.[14] But how the public would react to such an attack was an important factor. Admiral Radford later reflected that to defend the offshore islands against a determined PRC attack "it would undoubtedly be necessary to use atomic weapons." Eisenhower would have had to "consider feeling generated throughout the world and in China too, particularly if many civilians were killed."[15] Because of the potential negatives of using atomic weapons, they were often, even then, thought of mainly as deterrents.

Atomic Bombs as Deterrents

Atomic bombs were in fact thought to be perfect deterrent weapons. One idea of how best to deter a PRC attack on the offshore islands was to "let it be known that the presently deployed U.S. fighter unit on Formosa has an atomic capability—as do the carrier task forces in the area—and that they will use them to repel attack." A second suggestion was to redeploy a fighter-bomber wing with "atomic capability to Korea and, perhaps, another to Formosa." If these were not sufficient, then "even more conclusive actions by the U.S." could include "practice bombings-up of bomber units in the Far East with tactical atomic weapons" or even the "test of a penetration [that is, of bunkers or runways] A-Bomb on some stand-by airfield, possibly in the Marianas' or other islands under U.S. control in the Eastern Pacific."[16]

One adviser suggested to Dulles that if the Communists took the offshore islands from the Nationalists, the United States could "from time to time, drop an A-bomb on them to neutralize them and give the CHICOMS no advantage by their capture." But Dulles thought this would "accomplish nothing but the killing of a number of harmless fishermen." It would also be a "considerable waste of armament."[17] That was a concern, in that "we cannot splurge our limited supply of atomic weapons without serious danger to the entire international balance of power; and therefore any use which is made of them must be very carefully planned and thought out."[18]

In addition, the use of large numbers of atomic bombs to defend the islands could backfire, since it might involve the "wholesale use of atomic weapons against the densely populated mainland where land bursts would be required which would have a [radioactive] fall-out which might involve heavy casualties."[19] There would also "be [a] risk of large civilian casualties through after-effects, and indeed the inhabitants of Quemoy, and even of Taiwan, might not be immune under certain atmospheric and wind conditions." Such a strategy, then, would not be in the "long-range interest of the Republic of China," since the use of atomic weapons might "destroy any hope of good will and future favorable reception of the Republic of China by the [mainland] Chinese people."[20] Furthermore, it "might alienate Asian opinion and ruin Chiang Kai-shek's hopes of ultimate welcome back to the mainland."[21]

Chinese leaders responded quickly to these American announcements. In February 1955, Mao Zedong warned the Finnish ambassador that "if the Americans atomic-bombed Shanghai or Peking, 'they' [meaning the Soviets] would retaliate by wiping out American cities, which would cause the replacement of the present leaders of the United States." The Finnish ambassador checked with the Soviet ambassador, who assured him that "if the Americans bombed the Chinese mainland, the Soviet Government would give the Chinese all possible support under the Sino-Soviet Agreement."[22]

There was enormous concern in Washington that Moscow would use atomic bombs to support Beijing. The use of the phrase "all possible support" certainly implied that the USSR might resort to nuclear weapons to back China. However, Eden warned Dulles that Russia's goal might be quite different: "Involving US in these islands will put US on weakest ground with its allies and public opinion generally . . . [and the] Russians would probably find this situation to their advantage."[23] In other words, embroiling the PRC and the United States in a no-win conflict over the offshore islands would potentially give the USSR increased leverage over both.

Tactical Nuclear Bombs and Missiles

As weapons development progressed, smaller atomic bombs were created that could be used in tactical warfare. In April 1955, the Chief of Staff of the Army, Ridgway, notified the chairman of the JCS, Twining, that if requested as many as seven of the Army's eight batteries of M31 Honest John missiles (truck-mounted, surface-to-surface, unguided "artillery rockets" with a fifteen-mile range, organized in batteries of four launchers each) could be transferred to Taiwan; one of these batteries would be diverted from Japan to Taiwan in June 1955, while six batteries in Europe could be redeployed to Taiwan. According to an appendix dated April 4, 1955, this would allocate 161 nuclear missiles to Taiwan. While the Honest John missile could be used with conventional warheads, it was primarily intended to use the Mark 7 bomb's nuclear components. According to Ridgway: "Honest John batteries would provide a significant contribution to the defense of the island of Formosa. Their most effective use would be with atomic warheads."[24] As Eisenhower reminded the Joint Chiefs in July 1955, however, the very "principle of having a considerable amount of dispersal" of missile batteries was to "limit the effects of surprise attack."[25]

Because any American use of atomic weapons might spur the USSR into responding, the United States began to develop tactical nuclear weapons (that is, of even shorter range). Admiral Felt later recounted that by the end of the 1950s new breakthroughs made many more military options available and that "at that time we had plans for use of tactical nuclear weapons."[26] During June 1957, President Eisenhower met with Drs. Ernest O. Lawrence, Mark M. Mills, and Edward Teller to talk about the development of "clean" nuclear weapons and "tactical" fusion weapons. According to Teller, clean weapons could be "easily packaged," their effects consisting "only in the damage sought, i.e., only in the area of initial effects, free of fall-out outside this area." Lawrence emphasized that the development of "clean" nuclear weapons, as opposed to "dirty" bombs with widespread radioactivity effects, would be necessary to make such weapons useful on the battlefield, and "our failure could truly be a 'crime against humanity.'" In any case, "after clean weapons have been developed, it is possible to put 'additive materials' with them to produce radioactive fall-out if desired."[27]

In May 1958, the NSC examined the use of nuclear weapons in a conflict with China. The report assumed that the president would give permission but also that "nuclear weapons will be employed with the greatest selectivity, initially against the invading forces and supporting facilities, then against other airfield complexes, ports, staging bases, communications networks and similar military targets." While the report assumed also that the USSR would come to China's assistance, it assessed that "Soviet support would not be expected to include provision of nuclear weapons, at least not at the commencement of hostilities."[28]

Many military officers during this period did not believe that tactical nuclear weapons, as opposed to larger A-bombs, would lead to a larger war. On August 14, 1958, General Twining told Eisenhower that if the Chinese communists attacked the offshore islands, the American response might include "small atomic weapons." When questioned, Twining said that the Joint Chiefs of Staff "did not expect atomic retaliation if we attacked the two or three airfields nearest the coast with atomic weapons."[29] On August 29, a draft operational plan was sent to the Pacific Fleet stating, "It is probable that initially only conventional weapons will be authorized, but prepare to use atomic weapons. Prepare, if the use of atomic weapons is authorized, to extending bombing of ChiCom targets deeper into China as required."[30]

Also on the 29th, the president met with Acting Secretary of State Herter, Allen Dulles, Dulles's deputy Gen. Charles P. Cabell, Deputy Secretary of Defense Donald A. Quarles, General Twining, and Admiral Burke, among others, to formulate guidance for CinCPac. Three possible enemy scenarios were outlined: (a) simple harassment and interdiction, (b) an attack on one or more offshore islands, and (c) an attack "extending operations against Taiwan." It was decided to suspend the atomic decision for the present:

> With regard to phase *a,* Mr. Quarles suggested that our action broadly should be to support the Chinese Nationalists but try to stay out of the battle ourselves. In case phase *b* develops, we should authorize our commander to join the battle, but not to use atomic weapons nor to extend the area of combat beyond the immediate tactical area, including air fields. In case phase *c* develops, we would expect our commander to seek further instructions for more extended action. The President said he retains some question as to whether we should not authorize tactical atomic weapons in case of phase *b.* However, we cannot be sure this would be necessary, and since we do not want to outrage world opinion, perhaps we had better reserve this.

It was settled, however, that even in scenario *a,* the Chinese Nationalists could undertake "hot pursuit" against the enemy and that U.S. Navy "escort and protection operations could be authorized to the extent deemed militarily necessary and beyond Chinat capabilities, but confined to international waters."[31]

Control over use of atomic bombs rested with the president. The final message to Admiral Stump had a note at the end: "It was emphasized that during all three phases, atomic weapons could not be used until after specific authority has been obtained from the President."[32] Asked by a reporter whether he could assure the public that local commanders would really ask Washington "for permission to use such weapons which are available" and if so whether they would wait to "receive a reply or approval," Dulles responded, "I certainly can. The orders are very, very strict with regard to that."[33]

Also on the 29th, another message to CinCPac was circulated internally, in draft. It clearly stated the U.S. government's intent: "The United States Government will not permit the loss of the offshore islands to Chinese Communist aggression. In case of major air or amphibious attacks, which in the opinion of the United States seriously endanger the islands, the United States will concur in the ChiNat attack of ChiCom close-in mainland bases. In such an event, the United States will reinforce the ChiNats to the extent necessary to assure the security of these islands. This action may include joining in the attack of ChiCom bases, with atomic weapons used if needed to gain the military objective."[34]

Anglo-American Discussions about the Use of Atomic Bombs

If American military leaders treated atomic bombs like any other weapon, many civilian leaders in Washington were not as optimistic. Perhaps Dulles considered the question moot; as he told Prime Minister Macmillan in September 1958, "It seems that the Sino-Soviet strategy is designed to put strains upon us at many separate places and our various commitments to N.A.T.O., in Korea, to individual allies, are spreading our forces too thin for comfort—certainly unless atomic weapons are to be used."[35] Nevertheless, later that month Eisenhower assured Selwyn Lloyd that nuclear weapons would be used only for an "all-out effort rather than a local effort" and "that he did not plan to use nuclear weapons in any local situation at the present time."[36] Owing to fears that first use of atomic weapons could lead to reprisals, however, Admiral Felt was eventually directed to draw up a plan to defend Taiwan using only conventional weapons, which took some doing.[37]

Ongoing American-British talks on the use of nuclear weapons were particularly important, especially since Eisenhower thought a bit differently from his advisers, who were against using these weapons. Ike informed Churchill in Bermuda in 1953 that if the Communists orchestrated another major attack in Korea, "we intended to use every weapon in the bag, including our atomic types." With regard to the "use of atomic bomb in Korea in the event hostilities are initiated by the Reds," Eisenhower argued that the "atom bomb has to be treated just as another weapon in the arsenal," in particular since "anyone who held up too long in the use of his assets in atomic weapons might suddenly find himself subjected to such wide-spread and devastating attack that retaliation would be next to impossible."[38]

On September 2, 1958, General Twining explained to Dulles that a seven-to-ten-kiloton airburst would have a lethal range of three or four miles but that there would be "virtually no fall out." If tensions over the Taiwan Strait got out of hand, it might be necessary to use tactical weapons against the PRC: "The initial attack would be only on five coast airfields (with one bomb being used per airfield)."[39] However, a different report from the JCS on the same day envisioned using ten-to-eighty-kiloton weapons and "for air burst, not ground burst," which suggested higher levels of fallout. The use of atomic weapons against the mainland, as the second report argued, would be a necessity, since "conventional weapons would not be adequate to accomplish the elimination of the [PRC] installations."[40]

Assessment of all the factors involved convinced the secretary of state of the importance of a determined resistance: "If the Chicoms believe the US would actively intervene to throw back an assault, perhaps using nuclear weapons, it is probable there would be no attempt to take Quemoy by assault and the situation might quiet down, as in 1955." While there would be "strong popular revulsion against the US in most of the world," especially in Asia and "particularly harmful to us in Japan," there was hope that if "relatively small detonations were used with only air bursts, so that there would be no appreciable fallout or large civilian casualties, and if the matter were quickly closed, the revulsion might not be long-lived."[41]

During January 1958, Vice Adm. Austin K. Doyle, now Commander, U.S. Taiwan Defense Command, also reported that B61 Matadors (surface-to-surface, radio-guided, seven-hundred-mile cruise missiles) had been stationed in Taiwan and were "now set up ready for action if trouble should start." Although Doyle refused to say whether any atomic weapons were in Taiwan, it was public knowledge by this time that the Matador missiles were capable of delivering nuclear payloads in the forty-to-fifty-kiloton range. During 1958, the U.S. Navy also began to deploy the Mark 101 Lulu nuclear depth bomb. With its eleven-kiloton payload, the weapon was intended to destroy deeply submerged submarines.[42]

Conclusions

American views of atomic warfare changed during the 1950s from one of immediate use of the A-bomb, to deterrence, to selective employment of tactical weapons. The U.S. Navy was operating in a highly sensitive part of Asia, where the nuclear weapons of China's ally, the USSR, represented a special strategic concern (as would, after the PRC exploded its own atomic bomb in 1964, those of China itself). U.S. warships conducting patrols had always to be wary of the PRC's intentions. The danger of a small clash growing into a nuclear exchange was ever present.

Without a doubt, U.S. nuclear policy had a direct impact on Taiwan. For example, according to declassified reports, the U.S. military stored numerous atomic bombs in Taiwan, and these weapons were not removed until the early 1970s.[43] Also, during the mid-1960s the Nationalists began their own nuclear weapons program. In 1976, under pressure from the American government, Taiwan agreed to dismantle its program (following the 1995–96 Taiwan Strait crisis, President Lee Teng-hui would propose reactivating it).

Although the nuclear aspect of U.S. policy toward the Taiwan Strait was always kept highly secret, it was clear from 1950 onward that atomic weapons, both large bombs and smaller tactical devices, were available in the region should they be needed to prevent a PRC invasion of Taiwan. Over time, however, conventional systems like the Sidewinder air-to-air missile were perceived as more useful and began to replace them, at least the tactical atomic weapons. Also, as the Sino-Soviet "monolith" began to crumble during the late 1950s, it became less likely that the Soviet Union would come to China's rescue.

Notes

1. Marolda, "U.S. Navy and the Chinese Civil War," p. 180.

2. NSC Talking Paper.

3. Marolda, "U.S. Navy and the Chinese Civil War," pp. 189–90.

4. Memorandum of Conversation, Notes on Meeting in JCS Conference Room, December 1, 1950, Top Secret, Dean G. Acheson Papers, box 68, HSTPL.

5. Memorandum of Conversation, December 27, 1950, Top Secret, Dean G. Acheson Papers, box 68, HSTPL.

6. CinCPacFlt Op-Plan 51-Z-55, Top Secret, January 1955, DDE AWF, Int. Series 10, Formosa Area, U.S. Mil. Ops. (3), DDEPL.

7. Ibid., annex A, "Atomic Operations" [emphasis in original].

8. Dulles to Eisenhower, February 21, 1955.

9. Memorandum of Conversation with the President, March 7, 1955, Top Secret, Dulles, J. F., WHM, box 3, Meet. Pres. 1955 (7), DDEPL.

10. Memorandum for the Record, March 11, 1955, Top Secret, WH Office, OSANSA, NSC Series, Briefing Notes Subseries, box 17, Taiwan and the Offshore Islands, U.S. Policy toward (1955–58), DDEPL.

11. Memorandum for the Record, Meeting in President's Office, March 11, 1955, Top Secret–Eyes Only, DDE Ann Whitman, Int. Series 9, Formosa Visit to CinCPac 1955 (1), DDEPL.

12. Memorandum for the President, March 15, 1955, Top Secret, DDE AWF, Int. Series 9, Formosa Visit 1955 (1), DDEPL [emphasis original].

13. Memorandum for the Record, March 18, 1955, p. 3, Top Secret, DDE AWF, Int. Series 9, Formosa Visit 1955 (2), DDEPL.

14. "First Taiwan Strait Crisis: Quemoy and Matsu Islands."

15. Robertson to Dulles, April 25, 1955, sec. 1, p. 3, sec. 2, p. 2.

16. Immediate Actions re the Formosa Situation, March 19, 1955, pp. 6–7.

17. Memorandum of Meeting Held in the Secretary's Office, March 28, 1955, p. 11, Top Secret, Dulles, J. F., WHM, box 2, WHM 1955, Formosa Straits (1), DDEPL.

18. Ibid., p. 7.

19. Memorandum of Meeting with the Senators, April 28, 1955.

20. Preliminary Draft of Possible Statement of Position for Communication to the Republic of China, April 7, 1955, pp. 10–11.

21. Memorandum of Meeting with the Senators, April 28, 1955.

22. From U.K. Embassy in Peking to Foreign Office, February 5, 1955, Secret, PREM 11/867, TNA.

23. Dulles, telegrams to Eisenhower, February 25, 1955.

24. M. B. Ridgway [Gen.], memorandum to Chairman, Joint Chiefs of Staff, April 4, 1955, Top Secret, DDE AWF, Int. Series 10, Formosa Area, U.S. Mil. Ops. (1), DDEPL.

25. Memorandum of Conference with the President, July 7, 1955, Top Secret, DDE Papers 6, ACW Diary July 1955 (5), DDEPL.

26. Felt, *Reminiscences,* p. 396.

27. Memorandum of Conference with the President, June 24, 1957, Restricted Data, DDE Diary 25, June 57 Diary Staff Memos, DDEPL.

28. U.S. and Allied Capabilities for Limited Military Operations to 1 July 1961, May 29, 1958, pp. B11–B12.

29. Memorandum of Conference with the President, August 14, 1958, pp. 1–3.

30. Draft Message to CinCPac, August 29, 1958, p. 1, DDE AWF, Int. Series 11, Formosa (3), DDEPL.

31. Memorandum of Conference with the President, August 29, 1958, pp. 1–2.

32. JCS WASH DC, message to CinCPac, August 29, 1958, pp. 1–3.

33. News Conference, U.S. Naval Base, Newport, RI, September 4, 1958, p. 5.

34. Draft Message to CinCPac, August 29, 1958, p. 3.

35. John Foster Dulles, letter to Harold Macmillan, September 5, 1958, Top Secret, CAB 21/3272, TNA.

36. Memorandum of Conversation between the President, Lloyd, and Caccia, September 21, 1958, pp. 1–6.

37. Felt, *Reminiscences,* p. 396.

38. Discussion in Bermuda between Eisenhower and Churchill, December 4, 1953, DDE Diary Series 9, Diary: Copies of DDE Personal 1953–54 (1), DDEPL.

39. "Memorandum of Conversation," September 2, 1958, in *China, 1958–1960,* p. 120.

40. Taiwan Straits: Issues Developed in Discussion with JCS, September 2, 1958, p. 2.

41. Summary, Estimate of Factors Involved in the Taiwan Straits Situation, September 4, 1958, pp. 2–5.

42. "USS *Randolph* and the Nuclear Diplomatic Incident," *Nuclear Information Project,* accessed December 13, 2010, www.nukestrat.com/dk/randolph.htm.

43. In 1974, all nuclear weapons were moved from Taiwan to Clark Air Base in the Philippines. "CINCPAC Command History, 1974," linked to at "Nuclear Strategy Project: Japan FOIA Documents," *The Nautilus Institute: Digital Library,* oldsite.nautilus.org/archives/library/security/foia/japanindex.html.

CHAPTER SEVEN

Using Taiwan to Undermine the Sino-Soviet Alliance

Kill two birds with one stone. (一石二鳥.)

Rather than resort to force in the Taiwan Strait, the U.S. government hoped to defuse military tension and to focus instead on longer-term goals, such as breaking up the Sino-Soviet alliance. As early as November 3, 1948, the CIA hypothesized that "when the issue of subservience to Moscow has become more immediate than that of 'US imperialism,' Chinese nationalism will prove stronger than international Communism."[1] A national intelligence estimate of September 10, 1952, identified three specific issues that might at some point undermine this alliance: "The Chinese Communists might make demands upon the USSR, or even take action, incompatible with long-range Soviet global interests"; "Frictions might arise because of Soviet inability or disinclination to supply capital equipment"; and resistance might emerge to "a Communist world dominated from Moscow."[2]

China seemed to be the weaker partner in the Sino-Soviet alliance; the Russians had surrounded themselves with friendly governments in the Baltics and Eastern Europe, while "in the Far East there was a more direct confrontation between the Communist and non-Communist world."[3] This was exemplified in the PRC-ROC standoff over the offshore islands, and in the determination of each Chinese government to reunify all of China under its respective authority. In December 1954, for example, Taiwan's Foreign Minister Yeh had vehemently opposed any "two-China" theory.[4] But during a February 10, 1955, conversation Dulles cautioned Yeh that the solution to the communist threat would take time. Instead of trying to "force Chinese unification by military means" the United States and Taiwan should focus on "the vulnerability of Communist regimes to economic and other pressures."[5]

In particular, Washington hoped to undermine the Sino-Soviet alliance with economic tools. As early as January 17, 1951, the NSC had suggested that making China more economically dependent on Russia "might strain their relations."[6] On May 4, 1951, it recommended specific measures: to "continue United States economic restrictions against China" and "persuade other nations to adopt similar positions" so as to "stimulate differences between the Peiping [Beijing] and Moscow regimes."[7] Dulles's goal was to "strain the Sino-Soviet alliance by compelling the Chinese to increase economic and military demands for Soviet support to the point where Moscow would be forced to drop Beijing."[8] The primary means for carrying out this ambitious policy were the multilateral export controls, which "continued or enhanced stresses and strains on the ties between Communist China and its Soviet partners."[9]

"All Feasible Overt and Covert Means"

On September 15, 1947, the CIA assessed that the goal of Soviet removal of industrial machinery from Manchuria, which had begun in 1946, was to make China a "deindustrialized source of food and raw materials for the Soviet Far East," which, of course, would help make China a "minimum military threat to the USSR."[10] The NSC urged in 1949 that the U.S. government, while avoiding the "appearance of intervention" in China, "should be alert to exploit through political and economic means any rifts between the Chinese Communists and the USSR."[11]

On October 13, 1948, the NSC submitted a highly prescient report entitled "United States Policy toward China." It argued that while the USSR desired to exploit China's natural resources,

> it is the political situation in China which must arouse the aggressive interest of the Kremlin. In the struggle for world domination—a struggle which the Kremlin pursues essentially through political action (even in civil war)—the allegiance of China's millions is worth striving for. That allegiance is worth struggling for if only to deny it to the free world. In positive terms, China is worth having because capture of it would represent an impressive political victory and, more practically, acquisition of a broad human glacis from which to mount a political offensive against the rest of East Asia.

The CIA wisely concluded that Stalin would want the Chinese communists to dominate some part of China. He would not want all of it: "China is too big, too populous."[12]

Stalin's desire for control, of course, completely ignored Mao Zedong's intention to unify the whole country under the Chinese Communist Party, which meant, the CIA argued, that the Soviet government had to oppose him: "Even Mao and his colleagues cannot be permitted eventually to acquire all of it [mainland China]—the temptation [to attempt to do so] might be too great for them, especially as they would have, in part, risen to power on the heady wine of nationalism. The Kremlin prefers, where possible, not to take chances in such matters."[13] The USSR was particularly concerned that Chinese nationalism had been exacerbated by its own predatory actions against China, including "obvious Kremlin cupidity in northern Manchuria, its extraterritorial activities in Sinkiang and the dispatch of the Soviet Ambassador with the Nationalist Foreign Office to Canton" in the far south. Thus, "the full force of nationalism remains to be released in Communist China."[14]

On January 5, 1950, Truman told Senators Knowland and H. A. Smith that he "felt the Soviets were going to encounter increasing difficulties by way of their program of subjugation" of China.[15] The obvious American policy choice would be the opposite of whatever the USSR wanted. Thus, a Sino-Soviet split could be accomplished in a number of ways, including using "all feasible overt and covert means, consistent with a policy of not being provocative of war, to create discontent and internal divisions within each of the Communist-dominated areas of the Far East, and to impair their relations with the Soviet Union and with each other, particularly by stimulating Sino-Soviet estrangement."[16] Dean Acheson advised Truman on November 17, 1949, that the best policy was to "detach" China from its "subservience to Moscow and over a period of time encourage those vigorous influences which might modify it." Truman thought "this was the correct analysis."[17]

Soviet Eyes Are Bigger Than Their Stomachs

Arguably the most important U.S. government policy, however, was the most counterintuitive: to let the Communists dominate all of mainland China. The reasoning was actually quite simple: "There appears to be no chance of a split within the Party or between the Party and the USSR until the time of Communist domination of China."[18] At that point it would be necessary to make China as dependent as possible on the Soviet Union. To that end, prior to Mao Zedong's visit to Moscow in early 1950, the U.S. government refused to recognize Beijing, which meant "the Chinese Communists cannot now play off one great power against another, since they have no non-Soviet allies at the moment."[19]

Truman did this on purpose. Having learned as a boy the classic Ozark saying about how one's eyes are bigger than one's stomach, Truman meant to push the Soviet Union and China closer together in order ultimately to break them apart. According to one CIA report, the first part of that plan was exactly what the Soviet government was trying to do: "In China, Soviet policy is currently directed toward creating a situation that will preclude any possibility of the Chinese Communists asserting their independence from the Soviet bloc. The USSR undoubtedly assumes that, since the Chinese Communist program depends on Soviet aid, the Peiping regime will not resist Soviet penetration."[20] The USSR was particularly interested in the "realignment of China's foreign trade with the Soviet area instead of China's traditional trading partners among the Western nations."[21]

By early 1950 the United States was considering using Taiwan's strategic position to exert economic pressure on the PRC. On January 5 Maj. Edward C. Spowart, United States Army Reserve, suggested to Truman that they could "turn the table" on Russia: "If we sent Generalissimo Chiang Kai Shek [*sic*] the necessary military equipment such as radar, tanks (preferably M-5a)[,] light bombers, fighter craft, and small arms and ammunition with which he can defend Formosa and continue his attacks on the mainland from Formosa and place a blockade around China proper, Communist China's only sources of supply would be from Russia; and when Russia failed to keep the civilian and military personnel supplied with food, clothing and other essential materials, Communism will have failed in China."[22]

On May 17, 1951, the NSC announced that the nation's top goals were to "detach China as an effective ally of the USSR and . . . [d]eny Formosa to any Chinese regime aligned with or dominated by the USSR."[23] According to Consul General Walter McConaughy in Hong Kong, the Korean War made the PRC "so thoroughly dependent on the USSR for weapons and ammunition for the Korean war . . . [that] the point has now been reached where additional dependence on the Soviets may hasten the day when the Chinese become disillusioned with Russian aid."[24] The NSC directed its Psychological Strategy Board to "place maximum strain on the Soviet structure of power, including the relationships between the USSR, its satellites, and Communist China."[25] The idea was to promote "conflict between Chinese national interest and Soviet imperialism."[26] On December 22, 1954, Dulles told Eisenhower that he was counting on "the traditional Chinese dislike of foreigners which was bound in the long run to impair relations with Russia."[27] Or as Ike's longtime Army friend William R. Gruber had put it, more succinctly, "Every nation that has ever tried to control China has mired down in the attempt."[28]

A major focus of the U.S. government was cutting China's imports of necessary goods. The Western commodities China's factories most needed included raw cotton ($75 million's worth in 1951 dollars came from outside the Soviet bloc); chemicals, dyes, and drugs ($30 million); crude rubber ($25 million); iron and steel products ($25 million); machinery, vehicles, and cereals ($14 million); petroleum products ($12 million); jute products ($10 million); and nonferrous metals and manufactures ($10 million). Buying these from the Soviet bloc would require "long and costly transport routes" that would "likely involve losses on the part of one or the other of the trading partners."[29] Such economic warfare "would make Communist China more dependent on the very limited Chinese rail facilities connecting with the USSR," which in turn would "hamper current industrial production, retard industrial development and might seriously limit China's ability to sustain large-scale military operations."[30]

Dulles explained to the Canadian cabinet during a visit to Ottawa that the PRC's goal was to expand its power throughout East Asia: "I felt that the Chinese Communists felt that they were strong enough to make a worthwhile try at driving US influence away from the entire offshore island chain from the Aleutians to New Zealand and becoming themselves dominant in that part of the world."[31] One method of diverting or even stopping them was to exploit the "present weaknesses" of both the Soviet Union and the People's Republic, particularly those that might undermine their alliance. The Soviet Union, for instance, was facing severe challenges, including "unreadiness to wage general war, conflicts over its foreign policy, trouble in its leadership[,] and [an] agricultural crisis"; how, leaders in the White House asked, might these factors be used to impact "Soviet willingness adequately to support [it] indirectly or to become directly involved in Chinese Communist military operations against the off-shore islands or Formosa?"[32]

As noted earlier, the Nationalist blockage of Chinese coastal shipping meant that much of the PRC's domestic north–south trade had to be carried by train and most of its international trade over the Trans-Siberian Railway. Some of that trade then passed through Russia to a number of friendly Eastern European countries, but "at the moment transportation facilities limit this possibility."[33] (China did still conduct international maritime trade, but largely in either foreign-registered vessels or in Hong Kong ships, which were—in theory at least—neutral.) That diversion resulted in four additional costs: "(a) excessive transport costs, (b) adverse terms of trade for China, (c) significant delays in the delivery of industrial goods and consequent difficulties in procurement and planning, and (d) restrictions on Communist China's ability to earn foreign exchange and to obtain credit." In such added transport costs alone, it was estimated that China spent $32 million in 1955 and the USSR from $53 million to $76 million, a total extra transport cost of approximately $100 million. When higher prices for imported goods were factored in, it was calculated that China had to spend $172 million in 1955 just to offset the trade controls. Although, admittedly, the sum total of these costs was "small," the total effect was huge, aside from the enormous strains on China's railway system and the Trans-Siberian Railway: "They have an important immediate impact on military-industrial expansion which increases internal economic pressures and restricts Communist China's ability to embark on new military ventures in the Far East."[34]

All this was on purpose. The result was that by 1957, fully "50 percent of China's foreign trade was with the USSR."[35] When Anthony Eden argued that trade restrictions on China should be the same as those on Russia, the U.S. side replied that "even though the Soviets might try to ship considerable quantities of supplies to China, a much longer route and higher expense was involved for the Communists, and consequently we did not see any advantage in relaxing controls in the East."[36]

The Impact of Sino-Soviet Relations on a Taiwan Strait Resolution

The relationship between Sino-Soviet diplomatic relations and the first Taiwan Strait crisis is too rarely considered. Following Stalin's death in 1953, the leadership of the USSR was in turmoil. The last thing his successors wanted was to be dragged into a new world war over the Taiwan Strait, especially one that might become nuclear. At the same time that the U.S. government was attempting to restrain Chiang Kai-shek, therefore, the USSR was actively reducing its military support for China. In fact, just as the first Taiwan Strait crisis reached its peak in early 1955, Moscow withdrew important defensive installations from its former naval base at Lüshun. This action made Manchuria vulnerable to attack from the sea by the U.S. Navy, which put pressure on Beijing to de-escalate its military operations in the Taiwan Strait.

On February 1, 1955, Eisenhower wrote a long letter to General Gruenther in Europe. The president made it clear that while U.S. policy included breaking up the Sino-Soviet alliance, it should not be achieved through war:

> I do not believe that Russia wants war at this time—in fact, I do not believe that if we became engaged in rather a bitter fight along the coast of China, Russia would want to intervene with her own forces. She would, of course, pour supplies into China in an effort to exhaust us and certainly would exploit the opportunity to separate us from our major allies. But I am convinced that Russia does not want, at this moment, to experiment with means of defense against the [atomic] bombing that we *could* conduct against her mainland. At the same time, I assume that Russia's treaty with Red China comprehends a true military alliance, which she would either have to repudiate or take the plunge. As a consequence of this kind of thinking, she would probably be in a considerable dilemma if we got into a real shooting war with China. It would not be an easy decision for the men in the Kremlin, in my opinion.

Eisenhower's goal, therefore, was to "do what we believe to be right" and by showing firmness undermine the enemy's constant attempts "to disrupt the solidarity of the free world's intentions to oppose their aggressive practices." [37]

Changes in Sino-Soviet relations were an important backdrop to the peaceful resolution of the first Taiwan Strait crisis. In Moscow, political infighting continued through until February 20, 1956, when during the Twentieth Party Congress Nikita Khrushchev, as first secretary of the Communist Party of the Soviet Union, delivered his "secret speech" denouncing Stalin. During those three years, Mao Zedong attempted to renegotiate some of the more onerous aspects of the 1950 Sino-Soviet Treaty, and he began to assert his own leadership over the international communist movement.

Even before Stalin's death in 1953, important alterations in the 1950 Sino-Soviet Treaty had been made, especially concerning the status of the Manchurian port cities of Lüshun and Dalian. According to a 1952 agreement, all Soviet military forces were scheduled to withdraw from these Manchurian ports by May 31, 1955. Moscow also agreed to transfer "the installations in the area of the Port Arthur [Lüshun] naval base to the

Government of the People's Republic of China." When Moscow agreed to return the various Manchurian railways and the Lüshun naval base without charge, Mao evidently thought that the USSR would also leave the large-caliber guns that were protecting the port. But at the last minute Khrushchev refused, demanding instead that China pay for the guns at full price: "These are very expensive weapons[;] we would be selling them at reduced prices."[38] Also, as noted earlier, the Soviet Union's decision in 1955 to strip the base of its main defensive weapons left China highly vulnerable, the Sino-Japanese War of 1894–95 and the Russo-Japanese War of 1904–1905 having shown that its ports could easily be reduced from the sea.

These Russian actions were, perhaps, purposeful. During May 1955, Dulles told Molotov during meetings in Vienna that "we had obtained from the Chinese Nationalists arrangements which we thought would enable us to influence the situation for peace from our side and he suggested that the Soviet Union could do the same with the Chinese Communists." That was so particularly because "the Chinese Communists were dependent upon Russia for various strategic supplies and planes and could not develop their plans without Russian support."[39] Khrushchev's decision to strip the port defenses was a clear sign that the USSR was unwilling to back an offensive to retake the offshore islands. He further hinted that the USSR's nuclear umbrella might not cover the Taiwan Strait.

With its northern ports now defenseless, the PRC had little choice but to back down from its aggressive stance far to the south. The Eisenhower administration realized immediately that Beijing's policies had changed. Prior to February 1955, the PRC attacks on the offshore islands had seemed clearly intended as the "first phase of an attack against Taiwan . . . by force." However, after the enactment of the Formosa Resolution on January 29 and the Senate's approval of the Mutual Defense Treaty on February 9, 1955, "Communist propaganda on this subject [fell] to its lowest point for the past year," apparently owing at least in part to the sobering effect of the "display of U.S. sea and air power in the area."[40] The immediate crisis appeared to be over. On April 23, 1955, Beijing even announced that it was willing to negotiate the status of Taiwan, and on May 1 the PRC halted the shelling of Jinmen Island. Three months later, on August 1, as a sign of goodwill, China released eleven captured American airmen, who had previously been sentenced to lengthy jail terms.[41]

One can only speculate on the impact of Sino-Soviet relations on Beijing's decision to halt the southern offensive against Taipei. Certainly, the firmness of north/south linkages had been shown before, in particular by the impact of Eisenhower's "unleashing Chiang" policy: it had spurred on the successful negotiation of the Korean armistice, and in turn had then freed the PLA to redeploy troops to the south. Now, lacking Soviet support in the north, Beijing called off the attack against the offshore islands. During the first Taiwan Strait crisis, the PRC's leaders had clearly felt that given the correlation of forces, they had little choice but to back down. Ultimately, China's coercive tactics against Taiwan did not succeed, in large part because the USSR "failed to come to the PRC's rescue when it was intimidated by the United States."[42]

Growing Sino-Soviet Economic Tension

Behind the scenes were several important Sino-Soviet diplomatic factors that contributed to the second cross-strait crisis, in 1958. For instance, the PRC's determination to break the Nationalist blockade ultimately arose from Sino-Soviet disagreement over the "Great Leap Forward," China's recently adopted economic plan. Second, Mao's decision to shell Jinmen in 1958 without first seeking Soviet approval can be portrayed as "a challenge not just to Taipei and Washington but to Moscow's domination of the international Communist movement as well."[43]

On the economic side, it has been shown how the U.S.-led economic embargo had caused the PRC's economic dependency on the USSR to grow rapidly throughout the 1950s. In 1950 the PRC had borrowed three hundred million U.S. dollars from the USSR, a sum clearly insufficient to solve China's economic problems. The PRC's intervention that fall in the Korean War led not only to huge military losses but to greater indebtedness to the USSR, since to "add insult to injury, Stalin . . . demanded payment from China for the Soviet military equipment he had sent to Korea."[44]

During August 1958, Mao initiated his second five-year plan—the "Great Leap Forward"—a program of rapid, forced industrialization and agricultural collectivization. The combined effects of the Nationalist blockade and the U.S. embargo forced Mao to rely mainly on imports from the USSR for support: "Imports from the Soviet Union rose by an astounding 70 per cent in 1958 and 1959."[45] But Beijing's constant demands on Moscow would entail a heavy cost. During the summer of 1958, for example, Mao requested Soviet-built nuclear submarines. When Khrushchev asked how China would pay for them, Mao responded that "China had unlimited supplies of food." Beginning in 1959—when the catastrophic Great Leap was producing the Great Famine, in which ultimately tens of millions starved—the PRC exported millions of tons of grain, worth an estimated U.S.$935 million, largely to fund its purchases from the USSR.[46]

To increase these foreign grain sales, however, the PRC needed to increase its maritime trade. To that end, Chinese officials attempted to exploit Anglo-American differences in trade policy. Also, Mao put pressure on Taiwan to give up its last offshore island bases, the "only part of the mainland" the Nationalists still controlled.[47] Chiang's goal of returning would be undermined, and the PRC could finally claim to have reunified all of mainland China.

All these foreign policy, economic, and political factors contributed to Mao's decision in August 1958 to attack Jinmen. According to secret reports given to President Eisenhower, morale in the PRC was "so bad that the regime has just cancelled the visas of all foreign newspapermen and ordered them out of the country, apparently to prevent reports of how bad the situation is."[48]

Archival documents prove that Truman's and Eisenhower's administrations both meant to separate Russia and China. On November 18, 1958, Ike discussed with Dulles "our policy of holding firm until changes would occur within the Sino-Soviet bloc." It was neither a quick nor easy goal to achieve: "He [the president] felt that these [changes] were inevitable but realized that the policy we were following might not be popular. There were some who wanted to give in; others who wanted to attack. The policy that required patience was rarely popular."[49]

Conclusions

Faced with the prospect of "losing" China, American policy was to let the Russians have all the mainland. As early as January 2, 1950, Truman was told that the Chinese would never follow the "Moscow line": the "Chinese are not built that way." Furthermore, "Asia for the Asiatics" feelings had long been potent in China, and the Russians were "*not* accepted by Orientals as Asiatics."[50] In 1957, former French premier Edgar Faure, visiting Beijing, found that the U.S. policy of isolating the PRC would "increase China's extreme dependency on the USSR."[51] Indeed, one of Dulles's prime strategic goals was to push the USSR and the PRC together so as to tear them apart. That political outcome was largely achieved.

In the immediate aftermath of the Sino-Soviet split, which occurred right after the 1958 Taiwan Strait crisis, the PRC's trade with the USSR began to decline, just as its trade with the West began to grow. Chinese grain trade with Canada and Australia helped offset the effects of the Great Famine. In one historian's view,

> China's dependence on Soviet assistance inevitably created heavy economic burdens on Moscow and could slow down Soviet development, thus making the Moscow-Beijing alliance quite costly. On the other hand, Sino-Soviet economic leverage placed the Kremlin in a politically favorable position from which to dictate relations within the alliance and influence the [Chinese Communist Party]'s domestic and foreign policies. This paradoxical situation turned out to be a major contributor to the collapse of the Soviet economic cooperation and the eventual deterioration of the alliance between the two Communist powers.[52]

The U.S. government carried out a highly secretive and complex policy, using a wide variety of military, economic, and political means, of driving China and the USSR together so as to heighten their mutual hostility. In the end, the "indirect and long-term effect" of U.S. policies such as the strategic embargo produced such stress in Beijing's economic relations with Moscow that by 1960 it had "led to the disintegration of the Sino-Soviet alliance."[53] Control over the offshore islands had been crucial to the ultimate success of these policies.

Notes

1. Central Intelligence Agency, Possible Developments in China, November 3, 1948, Secret, PHST, President's Secretary's File, box 178, P.S.F. Subject File, HSTPL.

2. National Intelligence Estimate, Relations between the Chinese Communist Regime and the USSR: Their Present Character and Probable Future Courses, September 10, 1952, Secret, PHST, President's Secretary's File, box 215, P.S.F. Intelligence File, HSTPL.

3. Summary of Remarks of . . . Dulles, March 18, 1955, p. 1.

4. Memorandum of Conference with the President, December 20, 1954.

5. JFDP, February 21, 1955, reel 210/211, 92802, Princeton Univ.

6. NSC, Statement of the Problem, January 17, 1951.

7. NSC, United States Objectives, Policies and Courses of Action in Asia, May 4, 1951, Top Secret, PHST, President's Secretary's File, box 183, P.S.F. Subject File, HSTPL.

8. Lorenz M. Lüthi, The Sino-Soviet Split: Cold War in the Communist World (Princeton, NJ: Princeton Univ. Press, 2008), p. 247.

9. Multilateral Trade Controls against Communist China: U.S. Position Supporting No Reduction, Draft, January 7, 1956, p. 3, Secret, WH Office, OSANSA, NSC Series, Policy Paper Subseries, box 12, NSC 5429/5 Policy toward the Far East (1), DDEPL.

10. Central Intelligence Agency, Implementation of Soviet Objectives in China, September 15, 1947, Secret, PHST, President's Secretary's File, box 216, P.S.F. Intelligence Reports, ORE 45, HSTPL.

11. NSC, Policies of the Government of the United States of America Relating to the National Security, vol. 2, 1949, Top Secret, PHST, President's Secretary's File, box 171, P.S.F. Subject File, HSTPL.

12. NSC, "United States Policy toward China," October 13, 1948, Secret, PHST, President's Secretary's File, box 178, P.S.F. Subject File, HSTPL.

13. Ibid.

14. NSC, U.S. Policy toward China, February 28, 1949, Top Secret, PHST, President's Secretary's File, box 179, P.S.F. Subject File, HSTPL.

15. Memorandum of Conversation, Formosa Problem, January 5, 1950, Confidential, Dean G. Acheson Papers, box 66, HSTPL.

16. NSC 5429/5, "Policy toward the Far East," December 22, 1954, p. 13, Top Secret, WH Office, OSANSA, NSC Series, Policy Paper Subseries, box 12, DDEPL.

17. Conversation with the President, China and the Far East, November 17, 1949, Dean G. Acheson Papers, Memorandum of Conversation File, box 66, HSTPL.

18. Central Intelligence Agency, Chinese Communist Capabilities for Control of All China, December 10, 1948, Secret, PHST, President's Secretary's File, box 217, P.S.F. Intelligence Reports, ORE 77-48, HSTPL.

19. Summary of Telegrams, November 27, 1949, Top Secret, PHST, SMOF–Naval Aide, box 23, State Department Briefs File, HSTPL.

20. Ibid.

21. Central Intelligence Agency, Review of the World Situation, June 14, 1950, Secret, PHST, President's Secretary's File, box 181, P.S.F. Subject File, CIA 6-50, HSTPL.

22. Spowart to Truman, January 5, 1950.

23. NSC, United States Objectives, May 17, 1951.

24. Summary of Telegrams, September 18, 1951, Top Secret, PHST, SMOF–Naval Aide, box 24, State Department Briefs File, HSTPL.

25. NSC, Scope and Pace of Covert Operations, August 22, 1951, Top Secret–Eyes Only, PHST, President's Secretary's File, box 169, P.S.F. Subject File, HSTPL.

26. Central Intelligence Agency, Memorandum for the Senior NSC Staff, January 11, 1951, Secret, PHST, President's Secretary's File, box 173, P.S.F. Subject File, HSTPL.

27. Memorandum of Conversation with the President, December 22, 1954, Top Secret, Dulles, J. F., WHM, box 1, Meetings with the President, 1954 (1), DDEPL.

28. William R. Gruber to Harry H. Vaughan [Maj. Gen.], November 14, 1948, PHST, Official File, OF 150, box 759, File O.F. 150 Misc. (1947–48) [2 of 2], HSTPL.

29. Department of State, Office of Intelligence Research, Vulnerability of the Soviet Bloc to Existing and Tightened Western Economic Controls, January 26, 1951, Secret, PHST, President's Secretary's File, box 183, P.S.F. Subject File, HSTPL.

30. Central Intelligence Agency, Vulnerability of the Soviet Bloc to Economic Warfare, February 19, 1951, Secret, PHST, President's Secretary's File, box 183, P.S.F. Subject File, HSTPL.

31. Summary of Remarks of . . . Dulles, March 18, 1955, p. 1.

32. Immediate Actions re the Formosa Situation, March 19, 1955, pp. 3–4.

33. Report on U.S. Government Policies in Relation to China, n.d., p. 7.

34. Multilateral Trade Controls against Communist China, January 7, 1956, pp. 3, 7.

35. See Bruce A. Elleman and Stephen Kotkin, eds., *Manchurian Railways and the Opening of China: An International History* (Armonk, NY: M. E. Sharpe, 2010), p. 199.

36. Eisenhower Diary, Eden Talks, February 8, 1956, DDE Diary Series 9, Diary Copies of DDE Personal 1955–56 (2), DDEPL.

37. Eisenhower to Gruenther, February 1, 1955 [emphasis in original].

38. Nikita Khrushchev, *Memoirs of Nikita Khrushchev,* ed. Sergei Khrushchev, vol. 3, *Statesman, 1953–1964* (University Park: Pennsylvania State Univ. Press, 2007), p. 434.

39. Conversation at Ambassador's Residence, Vienna, on May 14, 1955, May 17, 1955, p. 2, Top Secret, DDE Subject Series, box 70, State, Dept. of (May 1955), DDEPL.

40. Edgar Eisenhower's copy of "Formosa" sent to Maxwell M. Rabb, April 7, 1955, p. 6, Secret, Dulles, J. F., WHM, box 2, Offshore April–May 1955 (5), DDEPL. This copy is dated April 7, 1955, though Dulles's "Formosa" draft cited above is of the 8th.

41. "First Taiwan Strait Crisis: Quemoy and Matsu Islands."

42. John F. Copper, "The Origins of Conflict across the Taiwan Strait: The Problem of Differences in Perceptions," in *Across the Taiwan Strait: Mainland China, Taiwan, and the 1995–1996 Crisis,* ed. Suisheng Zhao (New York: Routledge, 1999), p. 49, citing Weiqun Gu, *Conflicts of Divided Nations: The Case of China and Korea* (Westport, CT: Praeger, 1995), pp. 26–28.

43. Chen Jian, *Mao's China & the Cold War* (Chapel Hill: Univ. of North Carolina Press, 2001), p. 179.

44. Dikötter, *Mao's Great Famine,* p. 5.

45. Ibid., pp. 73–83.

46. Jung Chang and Jon Halliday, *Mao: The Unknown Story* (New York: Knopf, 2005), p. 428.

47. Washington, DC, telegram to Foreign Office, September 5, 1958, Top Secret, CAB 21/3272, TNA.

48. Memorandum of Conversations with the President, September 11, 1958, pp. 1–3, Top Secret, JFD Papers, WHM 7, WH Meet J–D [July–December] 1958 (6), DDEPL.

49. Memorandum of Conversation with the President, November 18, 1958, pp. 1–23, Secret, JFD Papers, WHM, box 7, Meetings July–Dec (3), DDEPL.

50. Goodier to Truman, January 2, 1950 [emphasis original].

51. Faure Views on Red China, July 16, 1957, Confidential, DDE Diary 25, July 57, Staff Memos (1), DDEPL.

52. Shu Guang Zhang, *Economic Cold War: America's Embargo against China and the Sino-Soviet Alliance, 1949–1963* (Stanford, CA: Stanford Univ. Press, 2001), pp. 268–69.

53. Ibid.

Conclusion
The Offshore Islands' Strategic Significance during the Cold War

Blood is thicker than water. (血濃於水.)

Tensions during the 1950s between the People's Republic of China and Taiwan were focused mainly on offshore islands in or near the Taiwan Strait. The U.S. government was concerned that a spark in this critical theater might erupt into war, perhaps even global conflict between the United States and the Sino-Soviet "monolith." Washington usually sought to lessen tensions over the offshore islands through its diplomatic relations with Taipei. But at other times Washington actively sought to exert military pressure—by "unleashing Chiang"—to force Beijing to move military units. As Senator Alexander Smith told Secretary of State Dulles during April 1955, if Korea and Indochina were the two flanks, Taiwan was the center, and "we should keep open a threat to the center in order to protect the two flanks."[1] In this regard, the Taiwan Patrol Force acted much like a vernier light switch, allowing the U.S. Navy to "dial" up or down cross-strait tension to suit larger East Asia policy objectives.[2]

Truman's refusal to become involved in China's civil war saved the United States from a quagmire. On January 11, 1949, the National Security Council argued that while the "objective of the U.S. with respect to China is the eventual development by the Chinese themselves of a unified, stable and independent China friendly to the U.S.," its goal was "not likely to be accomplished by any apparent Chinese group or groups [in] the foreseeable future."[3] On September 26, 1952, Truman received a "fan" letter quoting an admission in a speech the previous day by Senator Herbert H. Lehman of New York, a considerably more liberal-leaning Democrat than Truman, that China "could have been saved, if at all, [only] by all-out military intervention on our part." Not only would the United States have "become bogged down in the immense expanse of China, in a war to keep Chiang-Kai-Shek in power, [but it] . . . would have cost us millions of men and billions of dollars."[4]

American diplomacy with Chiang Kai-shek proved to be one of the greatest challenges. As the Department of State warned on January 19, 1949, when working with Taiwan the "choice is not between satisfactory and unsatisfactory courses of action but rather of the least of several evils or an amalgam of the lesser of them."[5] Two U.S.-Taiwanese secret agreements, one signed in December 1954 and the other in October 1958, allowed Washington to exert crucial leverage over Chiang to minimize the chances that a small conflict in the Taiwan Strait might accidentally escalate. By keeping the agreements secret Chiang could retain "face" even while ceding actual authority to Washington.

But there was also a highly negative side to secrecy. Because the specific terms of these agreements were unknown to the public, Washington often received the lion's share of global blame for rising tensions in the Taiwan Strait, even though it was in fact often working behind the scenes to quell them. Secrecy also offered Chiang greater public leverage to obtain his goals. In the end, however, all this was a price that Eisenhower and Dulles decided they would willingly pay to keep the peace in East Asia, where U.S. military policy in the Taiwan Strait was critical to stability.

On October 12, 1950, a top-secret CIA study assessed that "without direct Soviet participation and given strong naval and air assistance by the US armed forces, the Chinese Nationalist defense forces are capable of holding Formosa against a determined Chinese Communist invasion."[6] In March 1956 the British intelligence services concluded that although the Communists were capable of launching a full-scale attack on the offshore islands, it is "highly improbable that they will conduct military operations of this magnitude as long as the Seventh Fleet remains in the area." Instead of invading, therefore, the PRC decided, incorrectly, that time was on its side, that "it would be pointless to fight for areas which they hope to acquire in due course through subversion and propaganda."[7]

The most important U.S. military goal was to protect Taiwan from invasion by the PRC. An equally important political goal, however, was reassuring America's other East Asian allies, including Japan, South Korea, the Philippines, and Australia, that the PRC could not expand onto the first island chain. As early as June 10, 1949, Chennault warned, *"A Communist Asia Means the Loss of the Pacific Islands"* and further that "the loss of China sets off a chain reaction that may take months or even several years to accomplish."[8] This was an early description of the dreaded "domino effect"—with the theory that the fall of even a small East Asian nation to communism would lead to the fall of the rest.

The Japanese were especially impacted by Russian and Chinese expansionism. In early 1950, Hollington K. Tong, the U.S.-educated ROC ambassador to Japan and the United States, argued, "As long as the National Government survives at Formosa, with its constant threat of invasion of the mainland, Russia will go slow in its aggressive action against Japan. It is important to American security in Japan thus to keep Russia off-base in China."[9] During 1955, a Japanese official in Taipei emphasized that it was "the strategic value of the island itself," dominating the sea-lanes from Japan to the south, and "keeping it from the Chinese Communists" that mattered most to Tokyo.[10] Eisenhower privately told his friend Lew Douglas, "It is true that our strategic situation would be seriously—possibly even fatally—damaged in the Western Pacific if we should lose Formosa to the Communists."[11]

During 1958, discussing upcoming talks with Chiang Kai-shek, Dulles told the British that if the Japanese thought the United States was weaker than China, "they would go over to the Chinese Communists just as quickly as they could."[12] However, owing to Japan's World War II experience with the A-bomb, using atomic weapons to prove American strength could be equally dangerous; Ambassador Douglas MacArthur II in Tokyo warned that the consequences of using atomic weapons "could range from denial by the Japanese Government of continued use of our bases in Japan, either for operations or logistical support, to a strike by Japanese labor on the bases, which would make them virtually inoperable."[13] This complex situation tied Washington's hands; it had to support Taipei militarily but try its best to avoid using atomic bombs to do so.

It was widely believed that the best interest of all Asian countries could be furthered by forming a "Pacific Pact," similar to the North Atlantic Treaty Organization in Europe. However, as Gen. Douglas MacArthur told Truman at their October 15, 1950, meeting at Wake Island, "due to lack of homogeneity of the Pacific nations" these countries did not want to cooperate but instead wanted "assurance of security from the United States."[14] America's Asian alliances and coalition partners looked to the United States to protect Taiwan. As Dulles freely admitted, aiding the Nationalists might end up involving the United States in a "civil dispute." But China was not the only case of a divided country. Dulles listed others as "Korea, Germany, Austria, Vietnam and Laos. . . ." According to Dulles: "If there were armed attacks from the Communist portions of these countries against the non-Communist portions, we would aid the latter and in fact did so in Korea."[15]

In January 1950, President Truman was warned privately: "If we lose Formosa we will have lost not only a perfect military base for Army operations, but the controlling link between China, the Philippines, Indo China, Siam, Burma and all of India; and it is a well known fact that the Philippines will go Communist within a short period after the fall of Formosa."[16] Letting the PRC take the offshore islands would eventually lead to Chinese pressure against Taiwan sufficient to "bring about a government which would eventually advocate union with Communist China and the elimination of US positions on the islands."[17]

Should this happen, Truman was told (in another evocation of the domino theory), it would "seriously jeopardize" the anticommunist island barrier composed of Japan, South Korea, Taiwan, the Philippines, Thailand, and Vietnam. Other Southeast Asian states, including Indonesia, Malaya, Cambodia, Laos, and Burma would "probably come fully under Communist influence." Finally, American bases on Okinawa would become "untenable" and Japan would "probably fall within the Sino-Soviet orbit." These losses would happen gradually, over many years, but "the consequences in the Far East would be even more far-reaching and catastrophic than those which followed when the United States allowed the Chinese mainland to be taken over by the Chinese Communists, aided and abetted by the Soviet Union."[18]

One method to dissuade the PRC from invading Taiwan was to provide the National-ists with a dependable supply of military equipment and training to defend themselves, though not sufficiently advanced to allow Taiwan to attack the mainland. As a result, according to a U.S. intelligence advisory committee report of April 1957, in the near term the "Nationalists are very unlikely to launch an invasion or, in the absence of Chinese Communist provocation, to initiate other major military action."[19] Because the Chinese on Taiwan were displaying "a degree of cooperation and intelligent ef-fort scarcely equaled elsewhere in American experience," this partnership proved to be a "strategic bargain" for Washington: the "creation of United States military bases has been rendered unnecessary by the size and effectiveness of the Chinese military establishment."[20] This smaller "footprint" helped to relieve tension. After a particularly bad incident involving American military personnel and local civilians in Taiwan, Eisenhower is recorded as having told Dulles that "we must have a very serious look at these Asiatic countries, and decide whether we can stay there. It does not seem wise, if they hate us so much."[21] On June 4, 1957, Eisenhower told Senator Alexander Wiley, a Wisconsin Republican, that stationing troops in foreign countries "should only be done when impossible to avoid it"; he was doing what he "could to get our troops out, but of course it would be risking the collapse of our position in the Far East if we were to pull out of Formosa and Korea."[22]

The chairman of the JCS posed Washington's dilemma in the following terms: "Are we to risk loss of American prestige and influence in [the] world, through loss of the Off-shore Islands occasioned by failure to exert a maximum defense; or are we to risk loss of prestige and influence, through limited use of nuclear weapons to hold the Islands." Given these two options, the JCS reached a "consensus that we should take the second risk" and told President Eisenhower, "If we were to decide not to, we would have to recast our whole philosophy of defense planning. If we do not decide to take the second risk, each succeeding crisis and its concomitant decisions will become increasingly difficult for us."[23]

Secretary Dulles's stance on the use of atomic weapons was clear: as he once told General Twining, "There was no use of having a lot of stuff and never being able to use it."[24] Dulles reminded the president that "we have geared our defenses to the use of these [atomic weapons] in case of hostilities of any size, and . . . if we will not use them when the chips are down because of adverse world opinion, we must revise our defense setup."[25] But the fact that the United States was committed to protecting the offshore islands and Taiwan did not mean it advocated an attack on the mainland. On June 25, 1962, Rusk, now secretary of state, explained to the British foreign secretary, Alec Douglas-Home, that "Chiang Kai-shek would have no United States support if he attempted to attack the mainland," and it was agreed that the British chargé d'affaires would tell PRC leaders that the "United States had done and were doing everything possible to restrain the Nationalists from provocative action."[26]

In the midst of a life-and-death Cold War struggle, appeasement of communism was not considered to be an option. This left only two alternatives: "In the fullness of time either the communists will cease to be communists, whether by revolution or evolution, or the United States and its friends will have to fight them, whether in global or local wars."[27] Truman was accused of having given up on China, thereby losing "to our country most of the fruits of the victory won by our fighting men in the Pacific theatre."[28] However, at this point evolution was clearly preferable to global warfare. Eisenhower argued that it was necessary to "wage the cold war in a militant, but reasonable, style whereby we appear to the people of the world as a better group to band with than the Communists." If that was done correctly, "we have got a real fighting chance of bringing this world around to the point where the Communist menace, if not eliminated, will be so minimized it cannot work."[29]

That was an extremely long-term policy, not one that brought immediate benefits, but the U.S.-sponsored strategic sanctions and the Nationalist blockade that followed from it were highly successful in forcing evolutionary change on the PRC. With American training, equipment, and financial backing, it was hoped that "the blockade of war goods, partially effective now [there were an estimated thousand foreign-ship arrivals per year], will grow more effective as the naval and air forces of Nationalist China are built up with U.S. aid."[30]

A prime goal of the economic sanctions and blockade was to break up the Sino-Soviet alliance. Chennault warned in 1949 that it might take "decades or centuries" to subvert the "Communist dynasty in China."[31] But Secretary of State Acheson told Governor

Thomas E. Dewey of New York in an April 10, 1950, telephone conversation that the only flicker of "hope in the Far East is to drive a wedge between Peking and Moscow," a policy that it would be "unwise" for Dewey ever to mention in a public speech.[32] As early as December 1954, an NSC paper, "Policy toward the Far East," stated that a major objective was "disruption of the Sino-Soviet alliance through actions designed to intensify existing and potential areas of conflict or divergence of interest between the USSR and Communist China."[33] In May 1955, Senator Mundt congratulated Eisenhower for duly keeping the "Communists guessing. . . . [I]t creates much uneasiness among Communist high echelons in Russia and China."[34]

On the surface, the American and British economic policies with the PRC differed dramatically, but the two nations in fact sought the same goal: splitting the Sino-Soviet alliance apart and bringing the PRC into the Western "camp." In 1955, Eden even told Eisenhower that "the British government would always hope to be on the American side in every quarrel."[35] Britain did not follow the U.S. lead in cutting China completely off from international trade, but only because London thought it better to leave the door open by conducting limited trade, just in case China decided to turn to the West. This difference on means caused significant friction in Anglo-American relations. In private communications, however, Selwyn Lloyd reassured John Foster Dulles in September 1958 that "your troubles are our troubles. . . . [I]s there any way in which we can help?"[36]

The Anglo-American "carrot-and-stick" diplomatic approach had the desired long-term impact on China, in the end contributing to Beijing's decision to turn more to the West. Eisenhower was a firm supporter of the Republic of China on Taiwan, but on April 17, 1955, he told Dulles "that in the long run, unless the unexpected happened, it might be necessary to accept the 'Two China' concept." The president "spoke of '5–10 or 12 years.'"[37] Dulles declared to Molotov a month later that there was no need to let the Taiwan Strait start a world war: "Surely the situation could continue another decade or longer if the alternative was the risk of war within a year."[38]

Clearly, Washington was in no rush to solve the problem. By the late 1950s, a decade after the creation of the PRC, tension between China and the Soviet Union had reached a breaking point. Mao Zedong's attempt to pay off China's enormous debt to the USSR had brought nationwide famine.[39] Some historians have argued that the resulting Sino-Soviet rift took Washington by "surprise" and that the U.S. government did not adopt policies to "expedite the estrangement."[40] But others have seen that Dulles "sought to split the Chinese and Russians by driving them ever closer together."[41] The Sino-Soviet monolith's collapse during the late 1950s represented the full achievement of Truman's and Eisenhower's strategic objectives. During the next decade, Mao Zedong gradually turned China farther from the Soviet Union and closer to the West. In 1969, fourteen years after Eisenhower's and Dulles's evolutionary statements, President Richard

Nixon—who had been Eisenhower's vice president—was to begin the process of recognizing the People's Republic of China.

Even while the PRC was facing economic collapse, Taiwan was, with American assistance, on a completely different developmental path. British officials reported during 1955 that "Formosa is prosperous (with United States aid) and the people, whose standard of living is much higher than that of most Oriental countries, seem contented."[42] Having eschewed a simple military solution to China's unification, Chiang Kai-shek prophetically told an Australian newspaper that Taiwan would focus on economic growth: "We shall continue to build up Taiwan as an example of what free men can do."[43] American support was absolutely crucial to what was often called the "Taiwan miracle." As Eisenhower put it, "The vigorous and skilled population on Taiwan, the record of growth in investment and output, [and] the very real potential for acceleration, offer a prospect for a convincing demonstration that under free institutions a pace and degree of achievement can eventually be obtained in excess of that resulting under totalitarianism."[44]

In December 1956, the importance of this miracle to the rest of East Asia was plain: "Taiwan is becoming a show window of the free world, which impresses friends and foes alike with the constancy and success of American policy."[45] The NSC's policy for Taiwan emphasized also the need for greater democracy to accompany the growing economy, for "an increasingly efficient Government of the Republic of China (GRC), evolving toward responsible representative government, capable of attracting growing support and allegiance from the people of mainland China and Taiwan, and serving as the focal point of the free Chinese alternative to Communism."[46] Successful examples such as Taiwan were called by the diplomat James W. Riddleberger the "reverse-domino effect." Once these "islands of development" were formed, they could "give assistance and inspiration in their turn to other underdeveloped countries which are further behind in the growth process." By 1960, Taiwan was at the stage where it could help other developing countries, specifically by extending "technical assistance to free Vietnam."[47]

Defeating global communism was Eisenhower's overarching challenge; the defense of Taiwan was one major component of it. As he told Senator Knowland on January 26, 1955, "Co-existence is the absence of killing each other. I don't say it is satisfactory or that I like it. It simply means you and I are not killing each other off." With regard to Formosa and China, the "most we could ever have would be dual recognition of Nationalist China and Red China," but "at present neither side would accept it." Eisenhower concluded fatalistically: "Time seems to be the final arbiter of history."[48]

Eisenhower spoke of his global view also with his friend Lew Douglas: "The central fact of today's life is that we are in a life-and-death struggle of ideologies. It is freedom against dictatorship; Communism against capitalism; concepts of human dignity against the materialistic dialectic." America's opponents in this life-and-death struggle were playing by new rules: "They have complete contempt for any of those concepts of honor, decency and integrity which must underlie any successful practice of international law and order as we have always understood it." Sometimes Eisenhower was pessimistic: "I have come to the conclusion that some of our traditional ideas of international sportsmanship are scarcely applicable in the morass in which the world now flounders."[49]

By 1958, the small offshore islands in the Taiwan Strait, most of which an average American could not have named or pointed to on a map, had helped to exacerbate Sino-Soviet relations to the point of a split. This outcome was no accident but the result of measured and carefully thought-out policies pursued by the Truman and Eisenhower administrations. Among the most important were the strategic embargo adopted by the Truman administration in 1950 and support of the ten-year Nationalist blockade, which collectively and indirectly produced enormous friction between China and the USSR. That friction and rising military and political tensions eventually resulted in the Sino-Soviet split. Harry S. Truman's successor, Dwight D. Eisenhower, is usually credited for bringing these policies to fruition, but a wide variety of documents prove that these Cold War policies were nonpartisan and that they were embraced and supported by both presidents, a lesson contemporary politicians might do well to heed.

Notes

1. Memorandum of Meeting with the Senators, April 28, 1955.

2. See Bruce A. Elleman, *High Seas Buffer: The Taiwan Patrol Force, 1950–1979,* Newport Paper 38 (Newport, RI: Naval War College Press, 2012).

3. NSC, United States Policy toward China, January 11, 1949, Top Secret, PHST, President's Secretary's File, box 178, P.S.F. Subject File, HSTPL.

4. Guy Carolin to Harry S. Truman, September 26, 1952, PHST, Official File, OF 150, box 760, File O.F. 150 Misc. (1951–53) [1 of 2], HSTPL.

5. U.S. State Dept., The Position of the United States with Respect to Formosa, January 19, 1949, Top Secret, PHST, President's Secretary's File, box 178, P.S.F. Subject File, HSTPL.

6. CIA Report "Background on Possible Items for Discussion on Wake Island," October 12, 1950, Top Secret, PHST, President's Secretary's File, box 208, P.S.F. Korean War File, Wake Island Talks [1 of 2], HSTPL.

7. JFDP, March 6, 1956, reel 212/213, 94964, Princeton Univ.

8. Chennault to Steelman, June 10, 1949 [emphasis original].

9. Tong to Truman, January 4, 1950.

10. JFDP, November 22, 1955, reel 212/213, 94355, Princeton Univ.

11. Eisenhower to Douglas, March 9, 1955.

12. Foreign Office to Washington, DC, October 22, 1958.

13. Taiwan Straits: Issues Developed in Discussion with JCS, September 2, 1958, p. 3.

14. Substance of Statements Made at Wake Island Conference, October 15, 1950, Top Secret, PHST, President's Secretary's File, box 208, P.S.F. Korean War File, Wake Island Talk: Conference–Statements, HSTPL.

15. Dulles to Douglas, March 19, 1955.

16. Spowart to Truman, January 5, 1950.

17. Summary, Estimate of Factors Involved in the Taiwan Straits Situation, September 4, 1958, pp. 3–4.

18. Ibid.

19. JFDP, April 5, 1957, reel 216, 97367, Princeton Univ.

20. Report on Foreign Economic Policy Discussions, December 18, 1956, p. 7.

21. Phone Calls, May 24, 1957, DDE Diary 23, May 57, Phone Calls, DDEPL.

22. Legislative Leadership Meeting, June 4, 1957, Supplementary Notes, Confidential, DDE Diary 24, June 57 Misc. (2), DDEPL.

23. Taiwan Straits: Issues Developed in Discussion with JCS, September 2, 1958, pp. 3–4.

24. Telephone Call to General Twining, September 2, 1958, Dulles, J. F., Tel. Conv., box 9, Aug.–Oct. 1958 (4), DDEPL.

25. Memorandum of Conference with the President, September 4, 1958, Secret, DDE 36, Staff Notes Sept. 58, DDEPL.

26. "Record of a Conversation between the Foreign Secretary and Mr. Dean Rusk on Monday, June 25, 1962," n.d., PREM 11/3738, TNA.

27. Report on Foreign Economic Policy Discussions, December 18, 1956, p. 3.

28. Bullitt to Truman, February 9, 1948.

29. Conversation between the President and Senator Styles Bridges, May 21, 1957, pp. 1, 7, DDE Diary 24, May 57 Misc. (2), DDEPL.

30. "China Blockade: How It Works—Ships by the U.S.—Sailors by Chiang Kai-shek," *U.S. News and World Report,* February 20, 1953.

31. Chennault to Steelman, June 10, 1949.

32. Memorandum of Conversation, Dean Acheson and Governor Dewey, April 10, 1950, Secret, Dean G. Acheson Papers, box 67, HSTPL.

33. NSC 5429/5, "Policy toward the Far East," December 22, 1954, p. 3.

34. Memorandum to the president regarding Senator Mundt's letter, May 12, 1955.

35. Notes Dictated by the President Regarding His Conversation with Sir Anthony Eden, July 19, 1955.

36. Lloyd to Dulles, September 11, 1958.

37. Memorandum of Conversation with the President, Augusta, Georgia, 17 April 1955, April 18, 1955, p. 2, Top Secret, Dulles, J. F., WHM 3, Meet. Pres. 1955 (5), DDEPL.

38. Conversation at Ambassador's Residence, Vienna, May 17, 1955, p. 2.

39. Lawrence C. Reardon, *The Reluctant Dragon: Crisis Cycles in Chinese Foreign Economic Policy* (Seattle: Univ. of Washington Press, 2002), pp. 103–104.

40. Zhang, *Economic Cold War,* p. 237.

41. Marilyn Young, *The Vietnam Wars, 1945–1990* (New York: HarperCollins, 1991), pp. 309–10.

42. U.K. Consulate, Tamsui, to Foreign Office, February 24, 1955, Confidential, PREM 11/879, TNA.

43. JFDP, August 29, 1956, reel 214/215, 96035, Princeton Univ.

44. Hearings before the Committee on Foreign Affairs House of Representatives, March 1, 1960, p. 1, DDE U.S. Council on For. Econ. Policy, Randall Series, Agency Subseries, box 2, Int. Cooperation Adm. (1), DDEPL.

45. Report on Foreign Economic Policy Discussions, December 18, 1956, p. 8.

46. NSC, U.S. Policy toward Taiwan and the Government of the Republic of China, October 4, 1957, p. 1.

47. Hearings before the Committee on Foreign Affairs House of Representatives, March 1, 1960, p. 3.

48. Eisenhower's Meeting with Senator Knowland, January 26, 1955, DDE Papers 4, ACW Diary, January 1955 (1), DDEPL.

49. Eisenhower to Douglas, March 29, 1955, pp. 1–2.

Selected Bibliography

ARCHIVES AND MANUSCRIPT COLLECTIONS

Dulles, John Foster. Papers. Princeton Univ., Microfilm Series.

Dwight D. Eisenhower Library, Abilene, Kansas.

Harry S. Truman Presidential Library and Museum, Independence, Missouri.

The National Archives, ADM 1, 116; CAB 21; DEFE 13; FO 371; PREM 8, 11; T 237, Kew, United Kingdom.

Naval History and Heritage Command Archives, Washington, DC.

ORAL HISTORIES

Anderson, George W., Jr. *Reminiscences of Admiral George W. Anderson, Jr.* Oral History 42, Naval Institute Oral History Program, Annapolis, MD [hereafter NIOHP].

Ansel, Walter C. W. *The Reminiscences of Rear Admiral Walter C. W. Ansel.* Oral History 74, NIOHP.

Beshany, Philip A. *The Reminiscences of Vice Admiral Philip A. Beshany.* Oral History 45, NIOHP.

Bucklew, Phil H. *Reminiscences of Captain Phil H. Bucklew.* Oral History 34, NIOHP.

Burke, Arleigh A. *Recollections of Admiral Arleigh A. Burke.* Oral History 64, NIOHP.

Felt, Harry Donald. *Reminiscences of Admiral Harry Donald Felt.* Oral History 138, NIOHP.

Frankel, Samuel B. *The Reminiscences of Rear Admiral Samuel B. Frankel.* Oral History 325, NIOHP.

Smoot, Vice Adm. Roland N. "As I Recall . . . The U.S. Taiwan Defense Command," U.S. Naval Institute *Proceedings* 110/9/979 (September 1984), pp. 56–59.

Stroop, Paul D. *The Reminiscences of Vice Admiral Paul D. Stroop.* Oral History 139, NIOHP.

PUBLISHED DOCUMENTS / GOVERNMENT DOCUMENTS

Phillips, Steven E., ed. *China, 1969–1972.* Foreign Relations of the United States [hereafter FRUS], 1969–1976, vol. 17. Washington, DC: U.S. Government Printing Office [hereafter GPO], 2006.

Prescott, Francis C., Herbert A. Fine, and Velma Hastings Cassidy, eds. *The Far East: China.* FRUS, 1949, vol. 9. Washington, DC: GPO, 1974.

Schwar, Harriet D., ed. *China.* FRUS, 1955–1957, vol. 2. Washington, DC: GPO, 1986.

———, ed. *China.* FRUS, 1958–1960, vol. 19. Washington, DC: GPO, 1996.

BOOKS AND DISSERTATIONS

Bell, Christopher M., and Bruce A. Elleman, eds. *Naval Mutinies of the Twentieth Century: An International Perspective.* London: Frank Cass, 2003.

Bouchard, Joseph F. *Command in Crisis: Four Case Studies.* New York: Columbia Univ. Press, 1991.

Bussert, James, and Bruce A. Elleman. *People's Liberation Army Navy (PLAN): Combat Systems Technology, 1949–2010.* Annapolis, MD: Naval Institute Press, 2011.

Chang, Gordon H. *Friends and Enemies: The United States, China, and the Soviet Union, 1948–1972.* Stanford, CA: Stanford Univ. Press, 1990.

Chang, Jung, and Jon Halliday. *Mao: The Unknown Story.* New York: Knopf, 2005.

Chen Jian. *Mao's China & the Cold War.* Chapel Hill: Univ. of North Carolina Press, 2001.

Chen Ming-tong. *The China Threat Crosses the Strait: Challenges and Strategies for Taiwan's National Security.* Translated by Kiel Downey. Taipei: Dong Fan Color Printing, 2007.

Chiu, Hungdah. *China and the Taiwan Issue.* New York: Frederick A. Praeger, 1979.

Christensen, Thomas J. *Useful Adversaries: Grand Strategy, Domestic Mobilization, and Sino-American Conflict, 1947–1958.* Princeton, NJ: Princeton Univ. Press, 1996.

Clough, Ralph N. *Island China.* Cambridge, MA: Harvard Univ. Press, 1978.

Cohen, Warren I., ed. *New Frontiers in American–East Asian Relations.* New York: Columbia Univ. Press, 1983.

Dikötter, Frank. *Mao's Great Famine: The History of China's Most Devastating Catastrophe, 1958–1962.* New York: Walker, 2010.

Durkin, Michael F. *Naval Quarantine: A New Addition to the Role of Sea Power.* Maxwell Air Force Base, AL: Air Univ., Air War College, 1964.

Elleman, Bruce A., *High Seas Buffer: The Taiwan Patrol Force, 1950–1979.* Newport Paper 38. Newport, RI: Naval War College Press, 2012.

———. *Modern Chinese Warfare, 1795–1989.* London: Routledge, 2001.

———. *Moscow and the Emergence of Communist Power in China, 1925–30: The Nanchang Uprising and the Birth of the Red Army.* London: Routledge, 2009.

Elleman, Bruce A., and Stephen Kotkin, eds. *Manchurian Railways and the Opening of China: An International History.* Armonk, NY: M. E. Sharpe, 2010.

Elleman, Bruce A., and S. C. M. Paine. *Modern China: Continuity and Change 1644 to the Present.* Upper Saddle River, NJ: Prentice Hall, 2010.

———, eds. *Naval Blockades and Seapower: Strategies and Counter-strategies, 1805–2005.* London: Routledge, 2006.

———, eds. *Naval Power and Expeditionary Warfare: Peripheral Campaigns and New Theatres of Naval Warfare.* London: Routledge, 2011.

Gallagher, Rick M. *The Taiwan Strait Crisis.* Research Report 10-97. Newport, RI: Naval War College, Strategic Research Department, 1997.

Garver, John W. *China's Decision for Rapprochement with the United States, 1968–1971.* Boulder, CO: Westview, 1982.

———. *The Sino-American Alliance: Nationalist China and American Cold War Strategy in Asia.* Armonk, NY: M. E. Sharpe, 1997.

Gibert, Stephen P., and William M. Carpenter. *America and Island China: A Documentary History.* Lanham, MD: Univ. Press of America, 1989.

Glass, Sheppard. "Some Aspects of Formosa's Economic Growth." In *Formosa Today,* edited by Mark Mancall. New York: Frederick A. Praeger, 1964.

Hattendorf, John B., and Bruce A. Elleman, eds. *Nineteen-Gun Salute: Case Studies of Operational, Strategic, and Diplomatic Naval Leadership during the 20th and Early 21st Centuries.* Newport, RI: Naval War College Press, 2010.

Hickey, Dennis Van Vranken. *United States–Taiwan Security Ties: From Cold War to Beyond Containment.* Westport, CT: Praeger, 1994.

Hinton, Harold C. *China's Turbulent Quest.* New York: Macmillan, 1972.

Holober, Frank. *Raiders of the China Coast: CIA Covert Operations during the Korean War.* Annapolis, MD: Naval Institute Press, 1999.

Hugill, Paul D. *The Continuing Utility of Naval Blockades in the Twenty-First Century.* Fort Leavenworth, KS: U.S. Army Command and General Staff College, 1998.

Khrushchev, Nikita. *Memoirs of Nikita Khrushchev.* Edited by Sergei Khrushchev. Vol. 3, *Statesman, 1953–1964.* University Park: Pennsylvania State Univ. Press, 2007.

Kissinger, Henry. *White House Years.* Boston: Little, Brown, 1979.

Lasater, Martin L., ed. *Beijing's Blockade Threat to Taiwan: A Heritage Roundtable.* Washington, DC: Heritage Foundation, 1986.

Lilley, James R., and Chuck Downs, eds. *Crisis in the Taiwan Strait.* Washington, DC: American Enterprise Institute for Public Policy Research, 1997.

Liu, Ta Jen. *U.S.-China Relations, 1784–1992.* Lanham, MD: Univ. Press of America, 1997.

Lo Jung-pang. *China as a Sea Power, 1127–1368: A Preliminary Survey of the Maritime Expansion and Naval Exploits of the Chinese People during the Southern Song and Yuan Periods.* Edited by Bruce A. Elleman. Singapore: National Univ. of Singapore Press; Hong Kong: Hong Kong Univ. Press, 2012.

Lüthi, Lorenz M. *The Sino-Soviet Split: Cold War in the Communist World.* Princeton, NJ: Princeton Univ. Press, 2008.

Mancall, Mark, ed. *Formosa Today.* New York: Frederick A. Praeger, 1964.

Marolda, Edward J. *The Approaching Storm: Conflict in Asia, 1945–1965.* Washington, DC: GPO, 2009.

———. *By Sea, Air, and Land: An Illustrated History of the U.S. Navy and the War in Southeast Asia.* Washington, DC: GPO, 1994.

———. "Confrontation in the Taiwan Straits." In *U.S. Navy: A Complete History,* edited by M. Hill Goodspeed. Washington, DC: Naval Historical Foundation, 2003.

———. "Hostilities along the China Coast during the Korean War." In *New Interpretations in Naval History,* edited by Robert W. Love Jr., Laurie Bogle, Brian VanDeMark, and Maochun Yu. Annapolis, MD: Naval Institute Press, 2001.

———. *A New Equation: Chinese Intervention into the Korean War—Proceedings of the Colloquium on Contemporary History.* Washington, DC: Naval Historical Center, 1991.

———. "The U.S. Navy and the Chinese Civil War, 1945–1952." PhD diss., George Washington Univ., Washington, DC, 1990.

———. "Wall of Steel: Sea Power and the Cold War in Asia." In *Maritime Power in the Twentieth Century,* edited by David Stevens. St. Leonards, VIC, Austral.: Allen & Unwin, 1998.

Marolda, Edward J., and Oscar P. Fitzgerald. *The United States Navy and the Vietnam Conflict.* Vol. 2, *From Military Assistance to Combat, 1959–1965.* Washington, DC: Naval Historical Center, 1986.

Muller, David. *China as a Maritime Power.* Boulder, CO: Westview, 1983.

Paine, S. C. M. *The Wars for Asia, 1911–1949.* New York: Cambridge Univ. Press, 2014.

Park, Chang-Kwoun. "Consequences of U.S. Naval Shows of Force, 1946–1989." PhD diss., Univ. of Missouri–Columbia, August 1995.

Reardon, Lawrence C. *The Reluctant Dragon: Crisis Cycles in Chinese Foreign Economic Policy.* Seattle: Univ. of Washington Press, 2002.

Ryan, Mark A., David M. Finkelstein, and Michael A. McDevitt, eds. *Chinese Warfighting: The PLA Experience since 1949.* Armonk, NY: M. E. Sharpe, 2003.

Szonyi, Michael. *Cold War Island: Quemoy on the Front Line.* New York: Cambridge Univ. Press, 2008.

Tkacik, John J., Jr., ed. *Reshaping the Taiwan Strait.* Washington, DC: Heritage Foundation, 2007.

Tucker, Nancy Bernkopf, ed. *Dangerous Strait: The U.S.-Taiwan-China Crisis.* New York: Columbia Univ. Press, 2005.

Wang, Gabe T. *China and the Taiwan Issue: Impending War at Taiwan Strait.* Lanham, MD: Univ. Press of America, 2006.

Yang Zhiben [杨志本], ed. *China Navy Encyclopedia* [中国海军百科全书]. Vol. 2. Beijing: Sea Tide Press [海潮出版社], 1998.

Young, Marilyn. *The Vietnam Wars, 1945–1990.* New York: HarperCollins, 1991.

Zhang, Shu Guang. *Economic Cold War: America's Embargo against China and the Sino-Soviet Alliance, 1949–1963.* Stanford, CA: Stanford Univ. Press, 2001.

Zhao, Suisheng, ed. *Across the Taiwan Strait: Mainland China, Taiwan, and the 1995–1996 Crisis.* New York: Routledge, 1999.

INTERNET SOURCES

Abrahamson, Sherman R. "Intelligence for Economic Defense." *Central Intelligence Agency: Library.* www.cia.gov/library/center-for-the-study-of-intelligence/kent-csi/vol8no2/html/v08i2a03p_0001.htm.

Bublitz, Bob. "To Speak of Many Things." *VQ Association News Letter,* Winter/Spring 2006. Available at www.centurum.com/vq/pdf/VQ%20Winter-Spring%202006%20News3.pdf.

"CINCPAC Command History, 1974." Linked to at "Nuclear Strategy Project: Japan FOIA Documents," *The Nautilus Institute: Digital Library,* oldsite.nautilus.org/archives/library/security/foia/japanindex.html.

"First Taiwan Strait Crisis: Quemoy and Matsu Islands." GlobalSecurity.org. N.d. Accessed December 14, 2010. www.globalsecurity.org/military/ops/quemoy_matsu.htm.

Kawashima, Shin. "Soviet-Taiwanese Relations during the Early Cold War." *Cold War International History Project,* September 23, 2009.

www.wilsoncenter.org/index.cfm?topic_id=
1409&categoryid=696A6529-A719-72C4
-F6A95DC41879865E&fuseaction=topics
.events_item_topics&event_id=545522.

Keng, Robert. "Republic of China F-86's in
Battle." *ARC Air: Aircraft Resource Center.*
Accessed March 22, 2011. www.aircraft
resourcecenter.com/Stories1/001-100/021
_TaiwanF-86_Keng/story021.htm.

Marolda, Edward J. "Invasion Patrol: The Sev-
enth Fleet in Chinese Waters." Paper, "A New
Equation: Chinese Intervention into the
Korean War," Colloquium on Contemporary
History 3, June 20, 1990. Available at www
.history.navy.mil/colloquia/cch3c.htm.

"ROCAF F-104 Retirement." TaiwanAirPower
.org. www.taiwanairpower.org/history/
f104ret.html.

Schnabel, James F. "The Relief of MacArthur."
Chap. 20 in *Policy and Direction: The First
Year.* United States Army in the Korean
War. CMH Pub 20-1-1. Washington, DC:
U.S. Army Center of Military History, 1991.
Available at www.history.army.mil/books/
pd-c-20.htm.

Simonton, Ben. "Leadership in Frustration: A
Sea Story." *Gather.* www.gather.com/view
Article.jsp?articleId=281474976757231.

"Taiwan Strait: 21 July 1995 to 23 March 1996."
GlobalSecurity.org. www.globalsecurity.org/
military/ops/taiwan_strait.htm.

"USS *Randolph* and the Nuclear Diplomatic
Incident." *Nuclear Information Project.*
Accessed December 13, 2010. www.nuke
strat.com/dk/randolph.htm.

About the Author

Bruce Allen Elleman is the William V. Pratt Professor of International History at the Naval War College, in Newport, Rhode Island. He received at the University of California, Berkeley, the bachelor of arts degree in 1982 and completed at Columbia University the master of arts and Harriman Institute Certificate in 1984, the master of philosophy in 1987, the East Asian Certificate in 1988, and the PhD in 1993. In addition, he completed the master of sciences at the London School of Economics in 1985 and the master of arts in national security and strategic studies (with distinction) at the Naval War College in 2004. His twenty-five books are

- *China's Naval Operations in the South China Sea: Evaluating Legal, Strategic and Military Factors* (Renaissance Books, 2018)

- *Seaborne Perils: Piracy, Maritime Crime, and Naval Terrorism in Africa, South Asia, and Southeast Asia* (Rowman & Littlefield, 2018)

- *International Competition in China, 1899–1991: The Rise, Fall, and Restoration of the Open Door Policy* (Routledge, 2015)

- *Navies and Soft Power: Historical Case Studies of Naval Power and the Nonuse of Military Force,* Newport Paper 42, coedited with S. C. M. Paine (Naval War College Press, 2015)

- *Taiwan Straits: Crisis in Asia and the Role of the U.S. Navy* (Rowman & Littlefield, 2015)

- *Commerce Raiding: Historical Case Studies, 1755–2009,* Newport Paper 40, coedited with S. C. M. Paine (Naval War College Press, 2013)

- *Beijing's Power and China's Borders: Twenty Neighbors in Asia,* coedited with Stephen Kotkin and Clive Schofield (M. E. Sharpe, 2013)

- *High Seas Buffer: The Taiwan Patrol Force, 1950–1979,* Newport Paper 38 (Naval War College Press, 2012)

- *China as a Sea Power, 1127–1368: A Preliminary Survey of the Maritime Expansion and Naval Exploits of the Chinese People during the Southern Song and Yuan Periods* (unpublished manuscript by Lo Jung-pang), edited (National University of Singapore Press; Hong Kong University Press, 2012)

- *People's Liberation Army Navy (PLAN): Combat Systems Technology, 1949–2010,* coauthored with James Bussert (Naval Institute Press, 2011)

- *Naval Power and Expeditionary Warfare: Peripheral Campaigns and New Theatres of Naval Warfare,* coedited with S. C. M. Paine (Routledge, 2011)

- *Nineteen-Gun Salute: Case Studies of Operational, Strategic, and Diplomatic Naval Leadership during the 20th and Early 21st Centuries,* coedited with John B. Hattendorf (Naval War College Press, 2010)

- *Piracy and Maritime Crime: Historical and Modern Case Studies,* Newport Paper 35, coedited with Andrew Forbes and David Rosenberg (Naval War College Press, 2010)

- *Manchurian Railways and the Opening of China: An International History,* coedited with Stephen Kotkin (M. E. Sharpe, 2010)

- *Modern China: Continuity and Change 1644 to the Present,* textbook, coauthored with S. C. M. Paine (Prentice Hall, 2010)

- *Moscow and the Emergence of Communist Power in China, 1925–30: The Nanchang Uprising and the Birth of the Red Army* (Routledge, 2009)

- *Naval Coalition Warfare: From the Napoleonic War to Operation Iraqi Freedom,* coedited with S. C. M. Paine (Routledge, 2008)

- *Waves of Hope: The U.S. Navy's Response to the Tsunami in Northern Indonesia,* Newport Paper 28 (Naval War College Press, 2007)

- *Japanese-American Civilian Prisoner Exchanges and Detention Camps, 1941–45* (Routledge, 2006)

- *Naval Blockades and Seapower: Strategies and Counter-strategies, 1805–2005,* coedited with S. C. M. Paine (Routledge, 2006)

- *Naval Mutinies of the Twentieth Century: An International Perspective,* coedited with Christopher Bell (Frank Cass, 2003)

- *Wilson and China: A Revised History of the Shandong Question* (M. E. Sharpe, 2002)

- *Modern Chinese Warfare, 1795–1989* (Routledge, 2001)

- *Mongolia in the Twentieth Century: Landlocked Cosmopolitan,* coedited with Stephen Kotkin (M. E. Sharpe, 1999)

- *Diplomacy and Deception: The Secret History of Sino-Soviet Diplomatic Relations, 1917–1927* (M. E. Sharpe, 1997)

Several of Dr. Elleman's books have been translated into foreign languages, including *Modern Chinese Warfare,* republished in Chinese as *Jindai Zhongguo de junshi yu zhanzheng* (Taipei: Elite, 2002); and *Naval Mutinies of the Twentieth Century* in Czech as *Námořní vzpoury ve dvacátém století: Mezinárodní souvislosti* (Prague: BBart, 2004).

Index

Numbers in **bold** indicate pages with illustrations

The Newport Papers

On Wargaming: How Wargames Have Shaped History and How They May Shape the Future, by Matthew B. Caffrey Jr. (no. 43, January 2019).

Navies and Soft Power: Historical Case Studies of Naval Power and the Nonuse of Military Force, edited by Bruce A. Elleman and S. C. M. Paine (no. 42, June 2015).

Writing to Think: The Intellectual Journey of a Naval Career, by Robert C. Rubel (no. 41, February 2014).

Commerce Raiding: Historical Case Studies, 1755–2009, edited by Bruce A. Elleman and S. C. M. Paine (no. 40, October 2013).

Influence without Boots on the Ground: Seaborne Crisis Response, by Larissa Forster (no. 39, January 2013).

High Seas Buffer: The Taiwan Patrol Force, 1950–1979, by Bruce A. Elleman (no. 38, April 2012).

Innovation in Carrier Aviation, by Thomas C. Hone, Norman Friedman, and Mark D. Mandeles (no. 37, August 2011).

Defeating the U-boat: Inventing Antisubmarine Warfare, by Jan S. Breemer (no. 36, August 2010).

Piracy and Maritime Crime: Historical and Modern Case Studies, edited by Bruce A. Elleman, Andrew Forbes, and David Rosenberg (no. 35, January 2010).

Somalia . . . From the Sea, by Gary Ohls (no. 34, July 2009).

U.S. Naval Strategy in the 1980s: Selected Documents, edited by John B. Hattendorf and Peter M. Swartz (no. 33, December 2008).

Major Naval Operations, by Milan Vego (no. 32, September 2008).

Perspectives on Maritime Strategy: Essays from the Americas, edited by Paul D. Taylor (no. 31, August 2008).

U.S. Naval Strategy in the 1970s: Selected Documents, edited by John B. Hattendorf (no. 30, September 2007).

Shaping the Security Environment, edited by Derek S. Reveron (no. 29, September 2007).

Waves of Hope: The U.S. Navy's Response to the Tsunami in Northern Indonesia, by Bruce A. Elleman (no. 28, February 2007).

U.S. Naval Strategy in the 1990s: Selected Documents, edited by John B. Hattendorf (no. 27, September 2006).

Reposturing the Force: U.S. Overseas Presence in the Twenty-First Century, edited by Carnes Lord (no. 26, February 2006).

The Regulation of International Coercion: Legal Authorities and Political Constraints, by James P. Terry (no. 25, October 2005).

Naval Power in the Twenty-First Century: A Naval War College Review *Reader,* edited by Peter Dombrowski (no. 24, July 2005).

The Atlantic Crises: Britain, Europe, and Parting from the United States, by William Hopkinson (no. 23, May 2005).

China's Nuclear Force Modernization, edited by Lyle J. Goldstein with Andrew S. Erickson (no. 22, April 2005).

Latin American Security Challenges: A Collaborative Inquiry from North and South, edited by Paul D. Taylor (no. 21, 2004).

Global War Game: Second Series, 1984–1988, by Robert Gile (no. 20, 2004).

The Evolution of the U.S. Navy's Maritime Strategy, 1977–1986, by John Hattendorf (no. 19, 2004).

Military Transformation and the Defense Industry after Next: The Defense Industrial Implications of Network-centric Warfare, by Peter J. Dombrowski, Eugene Gholz, and Andrew L. Ross (no. 18, 2003).

The Limits of Transformation: Officer Attitudes toward the Revolution in Military Affairs, by Thomas G. Mahnken and James R. FitzSimonds (no. 17, 2003).

The Third Battle: Innovation in the U.S. Navy's Silent Cold War Struggle with Soviet Submarines, by Owen R. Cote Jr. (no. 16, 2003).

International Law and Naval War: The Effect of Marine Safety and Pollution Conventions during International Armed Conflict, by Dr. Sonja Ann Jozef Boelaert-Suominen (no. 15, December 2000).

Theater Ballistic Missile Defense from the Sea: Issues for the Maritime Component Commander, by Commander Charles C. Swicker, U.S. Navy (no. 14, August 1998).

Sailing New Seas, by Admiral J. Paul Reason, U.S. Navy, with David G. Freymann (no. 13, March 1998).

What Color Helmet? Reforming Security Council Peacekeeping Mandates, by Myron H. Nordquist (no. 12, August 1997).

The International Legal Ramifications of United States Counter-proliferation Strategy: Problems and Prospects, by Frank Gibson Goldman (no. 11, April 1997).

Chaos Theory: The Essentials for Military Applications, by Major Glenn E. James, U.S. Air Force (no. 10, October 1996).

A Doctrine Reader: The Navies of the United States, Great Britain, France, Italy, and Spain, by James J. Tritten and Vice Admiral Luigi Donolo, Italian Navy (Retired) (no. 9, December 1995).

Physics and Metaphysics of Deterrence: The British Approach, by Myron A. Greenberg (no. 8, December 1994).

Mission in the East: The Building of an Army in a Democracy in the New German States, by Colonel Mark E. Victorson, U.S. Army (no. 7, June 1994).

The Burden of Trafalgar: Decisive Battle and Naval Strategic Expectations on the Eve of the First World War, by Jan S. Breemer (no. 6, October 1993).

Beyond Mahan: A Proposal for a U.S. Naval Strategy in the Twenty-First Century, by Colonel Gary W. Anderson, U.S. Marine Corps (no. 5, August 1993).

Global War Game: The First Five Years, by Bud Hay and Bob Gile (no. 4, June 1993).

The "New" Law of the Sea and the Law of Armed Conflict at Sea, by Horace B. Robertson Jr. (no. 3, October 1992).

Toward a Pax Universalis: A Historical Critique of the National Military Strategy for the 1990s, by Lieutenant Colonel Gary W. Anderson, U.S. Marine Corps (no. 2, April 1992).

"Are We Beasts?" Churchill and the Moral Question of World War II "Area Bombing," by Christopher C. Harmon (no. 1, December 1991).

Newport Papers are available online (Acrobat required) at www.usnwc.edu/ Publications/Naval-War-College-Press/.